Beschaffenheit der Manuskripte

Die Manuskripte werden photomechanisch vervielfältigt; sie müssen daher in sauberer Schreibmaschinenschrift mit ausreichend großer Type geschrieben sein. Handschriftliche Formeln bitte nur mit schwarzer Tusche eintragen. Notwendige Korrekturen sind bei dem bereits geschriebenen Text entweder durch Überkleben des alten Textes vorzunehmen oder aber müssen die zu korrigierenden Stellen mit weißem Korrekturlack abgedeckt werden. Die reproduktionsfähigen Abbildungen (in Originalgröße) sollen in den Text eingeklebt werden. Falls das Manuskript oder Teile desselben neu geschrieben werden müssen, ist der Verlag bereit, dem Autor bei Erscheinen seines Bandes einen angemessenen Betrag zu zahlen. Die Autoren erhalten 50 Freiexemplare.

Zur Erreichung eines möglichst optimalen Reproduktionsergebnisses ist es erwünscht, daß bei der vorgesehenen Verkleinerung der Manuskripte der Text auf einer Seite in der Breite möglichst 18 cm und in der Höhe 26,5 cm nicht überschreitet. Entsprechende Satzspiegelvordrucke werden vom Verlag gern auf Anforderung zur Verfügung gestellt.

Manuskripte, in englischer, deutscher oder französischer Sprache abgefaßt, sind einzureichen bei: Springer-Verlag, 6900 Heidelberg, Postfach 1780.

Cette série a pour but de donner des informations rapides, de niveau élevé, sur des développements récents en physique, aussi bien dans la recherche que dans l'enseignement supérieur. On prévoit de publier.

1. des versions préliminaires de travaux originaux et de monographies
2. des cours spéciaux portant sur un domaine nouveau ou sur des aspects nouveaux de domaines classiques
3. des rapports de séminaires
4. des conférences faites lors de congrès ou de colloques

En outre il est prévu de publier dans cette série, si la demande le justifie, des rapports de séminaires et des cours multicopiés ailleurs mais déjà épuisés.

Dans l'intérêt d'une diffusion rapide, les contributions auront souvent un caractère provisoire; le cas échéant, les démonstrations ne seront données que dans les grandes lignes. Les travaux présentés pourront également paraître ailleurs. Une réserve suffisante d'exemplaires sera toujours disponible. En permettant aux personnes intéressées d'être informées plus rapidement, les éditeurs Springer espèrent, par cette série de «prépublications», rendre d'appréciables services aux instituts de physique. Les annonces dans les revues spécialisées, les inscriptions aux catalogues et les copyrights rendront plus facile aux bibliothèques la tâche de réunir une documentation complète.

Présentation des manuscrits

Les manuscrits, étant reproduits par procédé photomécanique, doivent être soigneusement dactylographiés type assez grand. Il est recommandé d'écrire à l'encre de Chine noire les formules non dactylographiées. Les corrections nécessaires doivent être effectuées soit par collage du nouveau texte sur l'ancien soit en recouvrant les endroits à corriger par du vernis correcteur blanc. Les illustrations; en dimension originale, préparées pour reproduction sont à insérer dans le texte. S'il s'avère nécessaire d'écrire de nouveau le manuscrit, soit complètement, soit en partie, la maison d'édition se déclare prête à verser à l'auteur, lors de la parution du volume, le montant des frais correspondants. Les auteurs recoivent 50 exemplaires gratuits.

Pour obtenir une reproduction optimale il est désirable que le texte dactylographié sur une page ne dépasse pas 26,5 cm en hauteur et 18 cm en largeur. Sur demande la maison d'edition met à la disposition des auteurs du papier spécialement préparé.

Les manuscrits en anglais, allemand ou français peuvent être adressés à Springer-Verlag, 6900 Heidelberg, Postfach 1780.

Lecture Notes in Physics

Edited by J. Ehlers, Austin, K. Hepp, Zürich and
H. A. Weidenmüller, Heidelberg
Managing Editor: W. Beiglböck, Heidelberg

5

Manfred Schaaf
Sektion Physik der Universität München

The Reduction of the Product of Two Irreducible Unitary Representations of the Proper Orthochronous Quantummechanical Poincaré Group

Springer-Verlag Berlin Heidelberg GmbH
1970

ISBN 978-3-540-05194-7 ISBN 978-3-540-36395-8 (eBook)
DOI 10.1007/978-3-540-36395-8

Library of Congress Catalog Card Number 72-139677.

Originally published by Springer-Verlag Berlin • Heidelberg 1970

Title No. 3324

Offsetdruck: Julius Beltz, Weinheim/Bergstr.

Table of Contents

Introduction

The knowledge of symmetries of a physical problem permits a reduction of matrix elements of observables. A classical example is offered by the Wigner-Eckart theorem for observables which transform like tensors under spatial rotations. In the S-matrix theory of elementary particles the assumption of Poincaré invariance of the scattering operator leads to a generalized partial wave expansion of S-matrix elements (cf. JOOS [1]) which requires the solution of the reduction problem for product representations of the Poincaré group $\tilde{P}$ and the derivation of the corresponding Clebsch-Gordan coefficients. This problem was at first solved for the so-called physical representations of $\tilde{P}$ for particles with four-momenta lying in the interior (JOOS [1]) or on the boundary (MOUSSA, STORA [2]) of the light cone. In the last years, however, also the so-called unphysical representations belonging to lightlike, spacelike or vanishing momenta have gained increasing importance for the S-matrix theory of elementary particles. For example, in connection with the analyticity and crossing symmetry postulates the so-called crossed partial wave expansion of the two-particle scattering amplitude which uses the coupling of an incoming and an outgoing particle (TOLLER [3], HADJEOANNOU [4]) has become interesting. Then the position of total momentum is occupied by the momentum transfer which is spacelike in most cases and to which therefore belong the "unphysical" representations with imaginary mass of the Poincaré group. The connection between Regge pole theory and group theory which becomes visible therewith has been pointed out especially by JOOS [5]. From a different point of view the interest in the "unphysical" representations of $\tilde{P}$ has been enlivened by FEINBERGs speculations [6] on the existence of faster than light "tachyons". Today therefore the reduction problem is coming into view also for products of "unphysical" representations of $\tilde{P}$.

We call Poincaré group $\tilde{P}$ the quantummechanical inhomogeneous Lorentz group; it is the universal covering group of the relativistic space-time transformation group the ray representations of which after BARGMANN [7] can be obtained as vector representations of $\tilde{P}$.+) Throughout this paper only the connected piece of the Poincaré group is considered, so that we can omit the precise denomination "proper orthochronous". The representation theory of $\tilde{P}$ in its essential features was founded by WIGNER in his famous paper [8] from 1939. A general mathematical

+) In this paper solely continuous representations are considered.

theory, the theory of induced representations of locally compact groups was developed later by MACKEY [9] (cf. also MACKEYs review article [10]), as a natural application of which one could conceive subsequently the work of Wigner. While as yet the representation theory of the Poincaré group is treated only very scantily in the textbooks on quantum field theory, there exist several lecture notes and original papers (cf. for instance EMCH [11], GUILLOT, PETIT [12]) in which it is developed from Mackeys theory. We therefore sketch in Chapter 1 the construction of the irreducible unitary representations of $\tilde{P}$ only in form of recipes. In somewhat more details, however, we will treat the representations belonging to spacelike momenta, because these usually come off badly in the physical literature. The representation theory of the little group SU(1,1) (respectively of its isomorphic image SL(2,$\mathbb{R}$)) belonging to spacelike momenta was founded by BARGMANN [13]. We present it in Chapter 1 in a version that uses results from GELFAND, GRAEV and VILENKIN [14]. In Chapter 2 we calculate matrix elements of the little group representations relative to bases of the representation spaces on which the little group representations, when restricted to certain subgroups, become diagonal. These matrix elements, as functions on the little groups, in some generalized sense are bases of the Hilbert spaces of square-integrable functions on the little groups or on certain coset spaces of the little groups. The respective expansion theorems are derived also in Chapter 2. Some of the analytic properties of the little group matrix elements, as established in this derivation, could also be of interest for the discussion of the analyticity properties of the S-matrix elements. The expansion theorems are needed for the solution, presented in Chapter 3, of the reduction problem for the product of any two irreducible unitary representations of $\tilde{P}$ with nonvanishing momenta. While the reduction method of spin-orbit coupling, as used by JOOS [1], essentially was suited only for the product of two representations with momenta lying in the interior of the light cone, MOUSSA and STORA [2] from Mackeys theory arrived at a reduction method, leading to the so-called helicity coupling, which is applicable to all cases with the exception of the product of two representations with spacelike or vanishing momenta. Our solution which also comprehends the last cases was inspired by the method of MOUSSA and STORA [2], but somewhat more emphasizes a geometrical-imaginative construction. The reduction problem for the product of two irreducible unitary representations of $\tilde{P}$ at least one of which belongs to vanishing momentum is deferred to reduction problems for certain unitary representations of the little groups the solutions of which in most cases exist in the literature.

Acknowledgement. I like to express my thanks to Dr. H.J. Meister for his persistent readiness to extensive discussions to which I owe many suggestions. I am greatly indebted to Prof. Dr. F. Bopp for his steady and friendly furtherance. I thank Dr. V. Ernst for his reading of my English translation of the manuscript. He has eliminated many mistakes and suggested improved formulations.
The German version of this paper was completed in March, 1969.

1 The Irreducible Unitary Representations of the Poincaré Group $\tilde{P}$

The Poincaré group $\tilde{P}$ consists of the pairs (A,a) with A from the group SL(2,$\mathbb{C}$) of complex unimodular 2x2-matrices, and a from the real four-dimensional vector space $\mathbb{R}^4$. The group law is

$$(1.1) \qquad (A,a)(A',a') = (AA',a+\Lambda(A)a') .$$

Here Λ: SL(2,$\mathbb{C}$) $\longrightarrow L_+^\uparrow$ is the covering homomorphism from SL(2,$\mathbb{C}$) onto the proper orthochronous Lorentz group $L_+^\uparrow$ which operates on $\mathbb{R}^4$ in the usual way. Λ is explicitly given by

$$A \longrightarrow \Lambda(A): \Lambda(A)^\mu{}_\nu = \frac{1}{2}\, Sp(\tilde{\sigma}^\mu A \sigma_\nu A^\dagger) ,$$

$$(1.2) \qquad (\sigma_\mu): = (\mathbb{1}_2,\vec{\sigma}) \ , \ (\tilde{\sigma}_\mu): = (-\mathbb{1}_2,\vec{\sigma}) \ ,$$

$$\sigma^1: = \begin{pmatrix} 0 & 1 \\ 1 & 0 \end{pmatrix} , \ \sigma^2: = \begin{pmatrix} 0 & -i \\ i & 0 \end{pmatrix} , \ \sigma^3: = \begin{pmatrix} 1 & 0 \\ 0 & -1 \end{pmatrix} .$$

As (1.1) shows, $\tilde{P}$ is a semidirect product SL(2,$\mathbb{C}$) Ⓢ $\mathbb{R}^4$ with the operation Λ of SL(2,$\mathbb{C}$) on $\mathbb{R}^4$, $\check{SL}(2,\mathbb{C})$: = (SL(2,$\mathbb{C}$),0) being a subgroup and $\check{\mathbb{R}}^4: = (\mathbb{1}_2,\mathbb{R}^4)$ being an abelian invariant subgroup of $\tilde{P}$.

1.1 The Construction of the Irreducible Unitary Representations of $\tilde{P}$

The following steps, according to MACKEY [9], lead to the irreducible unitary representations of $\tilde{P}$:

I. We set up the character group of the abelian invariant subgroup $\check{\mathbb{R}}^4$. We have

$$(1.1.1) \qquad (\mathbb{1}_2,a) \longrightarrow \chi^p(a) = e^{ip\cdot a}, \ p\cdot a: = -p^0a^0 + \vec{p}\cdot\vec{a}, \quad p\in\mathbb{R}^4 \ , \qquad ^{+)}$$

and therefore the character group is the "momentum space" $\mathbb{R}^4$. The use of the indefinite scalar product p·a is arbitrary but facilitates the performance of the following steps.

II. We set up the orbits of the group SL(2,$\mathbb{C}$) on the character group $\mathbb{R}^4$, i.e. the classes in $\mathbb{R}^4$ under the following equivalence relation:

+) We use in this paper the metric defined by $(g_{\mu\nu}): = \begin{pmatrix} -1 & & & \\ & 1 & & \\ & & 1 & \\ & & & 1 \end{pmatrix}$.

p and p' belong to the same orbit $\Omega(p)$, if $A \in SL(2,\mathbb{C})$ exists with $\chi^{p'}(a) = \chi^{p}(\Lambda(A)^{-1}a)$ for every $a \in \mathbb{R}^4$. According to (1.1.1) we have $\chi^{p}(\Lambda(A)^{-1}a) = \chi^{\Lambda(A)p}(a)$ and therefore $\Omega(p) = L_+^{\uparrow}p$, i.e. the orbit of p is the "mass shell" on which it lies.

III. We characterize the partition of $\mathbb{R}^4$ into orbits by the choice of the following set Ω of representing "standard momenta" $\mathring{p}$:

$$\mathbb{R}^4 = \bigcup_{\mathring{p}\in\Omega} \Omega(\mathring{p}), \qquad \Omega = \Omega^+ \cup \Omega^- \cup \Omega^o \cup \Omega_o^+ \cup \Omega_o^- \cup \Omega_o^o ,$$

$$(1.1.2) \qquad \Omega^{\pm} := \{\pm m\, e_{(0)} : m > 0\} , \quad \Omega^{o} := \{n\, e_{(3)} : n > 0\} ,$$

$$\Omega_o^{\pm} := \{\pm(e_{(0)} + e_{(3)})\} , \quad \Omega_o^o := \{0\} .$$

Here $\{e_{(\mu)}\}$ is an "orthonormal" basis of $\mathbb{R}^4$:

$$(1.1.3) \qquad e_{(\mu)} \cdot e_{(\nu)} = g_{\mu\nu} , \quad e_{(\mu)} e^{(\mu)} = \mathbb{1}_4 .$$

IV. We set up the little group $G(\mathring{p})$ for each $\mathring{p} \in \Omega$, i.e. the subgroup of $SL(2,\mathbb{C})$ for which $\chi^{\mathring{p}}(\Lambda(A)^{-1}a) = \chi^{\mathring{p}}(a)$ for every $A \in G(\mathring{p})$ and $a \in \mathbb{R}^4$. The last condition according to (1.1.1) and (1.2) is equivalent to

$$(1.1.4) \qquad A \in G(\mathring{p}) \Longleftrightarrow A\, \mathring{p}\cdot\sigma\, A^{\dagger} = \Lambda(A)\mathring{p}\cdot\sigma = \mathring{p}\cdot\sigma .$$

Because of (1.1.2) there exist only the following four different little groups:

$$(1.1.5) \qquad \begin{aligned} &\mathring{p} \in \Omega^+ \cup \Omega^- : G(\mathring{p}) = SU(2) \quad ; A \in SU(2) \Longleftrightarrow A_{21} = -A_{12}^*, A_{22} = A_{11}^* ; \\ &\mathring{p} \in \Omega^o \quad : G(\mathring{p}) = SU(1,1) ; A \in SU(1,1) \Longleftrightarrow A_{21} = A_{12}^*, A_{22} = A_{11}^* ; \\ &\mathring{p} \in \Omega_o^+ \cup \Omega_o^- : G(\mathring{p}) = E(2) \quad ; A \in E(2) \Longleftrightarrow A_{21} = 0, A_{22} = A_{11}^* ; \\ &\mathring{p} \in \Omega_o^o \quad : G(\mathring{p}) = SL(2,\mathbb{C}) . \end{aligned}$$

SU(2), SU(1,1) and E(2) are the subgroups of $SL(2,\mathbb{C})$ which are unitary relative to the definite metric $\sigma_o = \mathbb{1}_2$, the indefinite metric $\sigma_3 = \begin{pmatrix} 1 & 0 \\ 0 & -1 \end{pmatrix}$ and the degenerate metric $\sigma_o + \sigma_3 = 2\begin{pmatrix} 1 & 0 \\ 0 & 0 \end{pmatrix}$, respectively. E(2) is isomorphic to a twofold covering group of the group of euclidean motions in the plane. In Chapter 3 we also need the little group belonging to the set of representants

$$\Omega^{o\prime} := \{n\, e_{(2)} : n > 0\} . \tag{1.1.6}$$

According to (1.1.4), $G(\mathring{p})$ in this case is the real subgroup $SL(2,\mathbb{R})$ of $SL(2,\mathbb{C})$, isomorphic to $SU(1,1)$:

$$\mathring{p} \in \Omega^{o\prime}: G(\mathring{p}) = SL(2,\mathbb{R});\ A \in SL(2,\mathbb{R}) \Longleftrightarrow A = A^* . \tag{1.1.7}$$

V. The left coset space $SL(2,\mathbb{C})/G(\mathring{p})$ is homeomorphic with $\Omega(\mathring{p})$. From the left coset corresponding to $p \in \Omega(\mathring{p})$ which consis ts of all $A \in$ $\in SL(2,\mathbb{C})$ with $\Lambda(A)\mathring{p} = p$ we take a representant $A(p)$. If we parametrize the mass shells belonging to $\mathring{p} \neq 0$ in the following way:

$$p = \begin{cases} \pm m(\cosh\chi\, e_{(0)} + \sinh\chi\, e(\vartheta,\varphi)), & 0 \leq \chi < \infty \text{ for } p \in \Omega(\pm m\, e_{(0)}) , \\ n(\sinh\chi\, e_{(0)} + \cosh\chi\, e(\vartheta,\varphi)), & -\infty < \chi < \infty \text{ for } p \in \Omega(n\, e_{(3)}) , \\ \pm e^{\chi}\, (e_{(0)} + e(\vartheta,\varphi)) & , -\infty < \chi < \infty \text{ for } p \in \Omega(\pm(e_{(0)} + e_{(3)})), \end{cases} \tag{1.1.8}$$

$$e(\vartheta,\varphi) := \sin\vartheta\,(\cos\varphi\, e_{(1)} + \sin\varphi\, e_{(2)}) + \cos\vartheta\, e_{(3)} ,\ 0 \leq \vartheta \leq \pi ,\ 0 \leq \varphi < 2\pi,$$

we may choose as the corresponding representants

$$A(p) = \hat{A}(\vartheta,\varphi)\hat{A}(\chi),\ \hat{A}(\vartheta,\varphi) := \begin{pmatrix} \cos\vartheta/2 & -e^{-i\varphi}\sin\vartheta/2 \\ e^{i\varphi}\sin\vartheta/2 & \cos\vartheta/2 \end{pmatrix},$$

$$\hat{A}(\chi) := \begin{pmatrix} e^{\chi/2} & 0 \\ 0 & e^{-\chi/2} \end{pmatrix} . \tag{1.1.9}$$

$\hat{A}(\chi)$ effects a pure velocity transformation in the $e_{(0)} - e_{(3)}$-plane with velocity $\mathrm{tgh}\chi$, $\hat{A}(\vartheta,\varphi)$ the shortest rotation from the $e_{(3)}$-direction into the direction of $\vec{p}$. The case $\mathring{p} = 0$ is trivial since $SL(2,\mathbb{C})/G(0)$ consists of one element only.

VI. For each little group we set up a system of representants of the equivalence classes of all irreducible unitary representations $\{U^{\rho}_{G(\mathring{p})}\}$. The index ρ may characterize the different classes. Let $\mathfrak{H}^{\rho}_{G(\mathring{p})}$ be the

Hilbert space on which $U^{\rho}_{G(\mathring{p})}$ operates. The explicit form of the system $\{U^{\rho}_{G(\mathring{p})}\}$ will be presented in Sections 1.2 - 1.5.

VII. For the subgroup $\check{G}(\mathring{p}) := G(\mathring{p}) \circledS \mathbb{R}^4 \subset \tilde{P}$ one can construct now on $\mathfrak{h}^{\rho}_{G(\mathring{p})}$ the representation

$$(1.1.10) \quad \check{G}(\mathring{p}) \ni (A,a) \longrightarrow U^{\mathring{p},\rho}_{\check{G}(\mathring{p})}(A,a) := \chi^{\mathring{p}}(a)\, U^{\rho}_{G(\mathring{p})}(A) \quad .$$

VIII. For $\mathring{p} \neq 0$ a measure on $\Omega(\mathring{p})$, invariant under SL(2,$\mathbb{C}$), is defined by

$$(1.1.11) \qquad \int_{\Omega(\mathring{p})} d\omega_{\mathring{p}}(p) := \int_{\Omega(\mathring{p})} \frac{d^3\vec{p}}{2|p^0|} \quad .$$

With this one sets up the direct integral

$$(1.1.12) \qquad \mathfrak{h}^{\mathring{p},\rho} := \oplus\int_{\Omega(\mathring{p})} \sqrt{d\omega_{\mathring{p}}(p)}\; \mathfrak{h}^{\rho}_{G(\mathring{p})}(p)$$

of the Hilbert spaces $\mathfrak{h}^{\rho}_{G(\mathring{p})}(p) \equiv \mathfrak{h}^{\rho}_{G(\mathring{p})}$. $\mathfrak{h}^{\mathring{p},\rho}$ consists of vector-valued functions $\psi: \Omega(\mathring{p}) \longrightarrow \mathfrak{h}^{\rho}_{G(\mathring{p})}$ with scalar product

$$(1.1.13) \qquad \langle\psi|\varphi\rangle^{\mathring{p},\rho} := \int_{\Omega(\mathring{p})} d\omega_{\mathring{p}}(p) \langle\psi(p)|\varphi(p)\rangle^{\rho}_{G(\mathring{p})} \quad ,$$

where $\langle\;|\;\rangle^{\rho}_{G(\mathring{p})}$ means the scalar product of $\mathfrak{h}^{\rho}_{G(\mathring{p})}$. For $\mathring{p} = 0$ we have $\mathfrak{h}^{0,\rho} \equiv \mathfrak{h}^{\rho}_{G(0)}$ with the scalar product

$$(1.1.14) \qquad \langle\psi|\varphi\rangle^{0,\rho} \equiv \langle\psi|\varphi\rangle^{\rho}_{G(0)} .$$

IX. On $\mathfrak{h}^{\mathring{p},\rho}$ the representation (1.1.10) of the subgroup $\check{G}(\mathring{p})$ for $\mathring{p} \neq 0$ induces the representation

$$(U^{\mathring{p},\rho}(A,a)\psi)(p) := U^{\mathring{p},\rho}_{\check{G}(\mathring{p})}(R(p;A),\Lambda(p)^{-1}a)\,\psi(\Lambda(A)^{-1}p) =$$

$$(1.1.15) \qquad = e^{ip\cdot a}\, U^{\rho}_{G(\mathring{p})}(R(p;A))\,\psi(\Lambda(A)^{-1}p) \quad ,$$

$$\Lambda(p) := \Lambda(A(p)),\; p = \Lambda(p)\mathring{p} \;,\; R(p;A) := A(p)^{-1}\, A\, A(\Lambda(A)^{-1}p)$$

of the Poincaré group [+)]. For $\mathring{p} = 0$ we have instead of (1.1.15) the

+) The choice of the representants A(p) in (1.1.8) and (1.1.9) here leads to the so-called helicity representations. For if $A = \exp(\frac{i}{2}\alpha\frac{\vec{p}\cdot\vec{\sigma}}{|\vec{p}|})$ is a rotation about the direction of momentum, $R(p;A) = \exp(\frac{i}{2}\alpha\sigma_3)$ is

representation

$$U^{0,\rho}(A,a)\,\psi = U^{\rho}_{G(0)}(A)\,\psi \tag{1.1.16}$$

of $\tilde{P}$ on $\mathfrak{H}^{\rho}_{G(0)}$. In it the translational normal subgroup $\check{\mathbb{R}}^4 \subset \tilde{P}$ is represented trivially.

From the general results of Mackey [9] one can be sure, that
a) every irreducible unitary represntation of $\tilde{P}$ is equivalent to one of the representations $U^{\mathring{p},\rho}$ constructed above,
b) two representations $U^{\mathring{p},\rho}$ and $U^{\mathring{p}',\rho'}$ arrived at in this way are equivalent if and only if $\mathring{p} = \mathring{p}'$ and $\rho = \rho'$.
Therefore the problem of finding all irreducible unitary representations of the Poincaré group $\tilde{P}$ is reduced to the problem of setting up all irreducible unitary representations of the little groups.

1.2 The Irreducible Unitary Representations of SU(2)

The representation theory of the group SU(2), the universal covering group of the group of rotations in three dimensions, is known to physicists since the early days of quantum mechanics. Referring to the review article [15] of BARGMANN we therefore may brief on this.

On the Hilbert space $\mathfrak{H}$ of entire analytic functions of two complex variables with the scalar product

$$\begin{aligned} \langle f|g\rangle :&= \int_{\mathbb{C}^2} d\mu(z_1,z_2)\, f(z_1,z_2)^* g(z_1,z_2)\ , \\ d\mu(z_1,z_2) :&= \frac{e^{-|z_1|^2-|z_2|^2}}{\pi^2}\, dx_1 dy_1 dx_2 dy_2,\ (z_1,z_2)=(x_1+iy_1,x_2+iy_2)\in\mathbb{C}^2, \end{aligned} \tag{1.2.1}$$

a unitary representation of SU(2) is defined by

$$\text{(1.2.2)}\quad SU(2) \ni A \longrightarrow U(A):\ (U(A)f)(z_1,z_2) = f(z_1A_{11}+z_2A_{21}, z_1A_{12}+z_2A_{22}).$$

We decompose $\mathfrak{H}$ into the direct sum

---

a rotation about the $e_{(3)}$-axis. If one chooses therefore, as usually, a basis for the representations of the little groups in which the rotations about the $e_{(3)}$-axis are diagonal, in the induced representation of $\tilde{P}$ the rotations about the direction of the momentum are diagonal; the helicity is the directional quantum number of angular momentum.

$$(1.2.3)\quad \mathfrak{H} = \bigoplus_{\substack{\varkappa=0,1\\ \ell=0,1,\ldots}} \mathfrak{H}^{\varkappa,\ell}_{SU(2)},\quad \mathfrak{H}^{\varkappa,\ell}_{SU(2)} := \{f\in\mathfrak{H}: f(az_1,az_2)=a^{2\ell+\varkappa}f(z_1,z_2)\}$$

of the finite-dimensional subspaces $\mathfrak{H}^{\varkappa,\ell}_{SU(2)}$, $\varkappa = 0,1$; $\ell = 0,1,2,\ldots$ of all homogeneous polynomials of degree $2\ell+\varkappa$. The representation U decomposes into the direct sum of the irreducible representations

$$(1.2.4)\quad SU(2)\ni A \longrightarrow U^{\varkappa,\ell}_{SU(2)}(A) := U(A)\big|\mathfrak{H}^{\varkappa,\ell}_{SU(2)},\quad \varkappa = 0,1;\ \ell = 0,1,2,\ldots .$$

Each irreducible unitary representation of SU(2) is equivalent to one of the representations $U^{\varkappa,\ell}_{SU(2)}$, $U^{\varkappa,\ell}_{SU(2)}$ and $U^{\varkappa',\ell'}_{SU(2)}$ being equivalent if and only if $\varkappa = \varkappa'$ and $\ell = \ell'$.

Customarily one uses instead of the pair $(\varkappa,\ell)$ the quantum number $j = \ell+\varkappa/2$ of the angular momentum which runs through the integer and the halfinteger numbers. We prefer the above notation because it allows the discrimination between the integer ($\varkappa = 0$) and the halfinteger ($\varkappa = 1$) spins.

1.3 The Irreducible Unitary Representations of SU(1,1)

The representation theory of the noncompact group SU(1,1) was founded in 1947 by BARGMANN [13]. Since then it has scarcely appeared in the physical literature. So we present it in somewhat more details than the representation theory of SU(2). The general frame will be marked by the book of GELFAND, GRAEV and VILENKIN [14], to which we refer also for the proof of some mathematical statements.

Call $\mathfrak{F}$ the linear space of functions defined on the boundary of the unit circle $\{\omega\in\mathbb{C}: |\omega| = 1\}$ and infinitly often differentiable there with the topology defined by pointwise convergence. It is known that every $f\in\mathfrak{F}$ can be expanded with respect to the basis

$$(1.3.1)\qquad \{e_\nu : e_\nu(\omega) = \omega^\nu,\ \nu = 0,\pm1,\pm2,\ldots\}$$

into an everywhere convergent Fourier series

$$(1.3.2)\qquad f = \sum_{\nu=-\infty}^{+\infty} f_\nu e_\nu,\quad f_\nu := \frac{1}{2\pi i}\oint\frac{d\omega}{\omega}\, e_\nu(\omega)^* f(\omega) .$$

Via the Fourier components we define the subspaces

$$\mathfrak{F}_{+}^{\varkappa,n} := \{ f \in \mathfrak{F} : f_\nu = 0 \text{ for } \nu < n+\varkappa+1 \} ,$$

(1.3.3)

$$\mathfrak{F}_{-}^{\varkappa,n} := \{ f \in \mathfrak{F} : f_\nu = 0 \text{ for } \nu > -n-1 \}$$

of $\mathfrak{F}$. From $\mathfrak{F}_{\pm}^{\varkappa,n}$ one obtaines the subspaces

$$\text{(1.3.4)} \qquad \mathfrak{F}_{\cap}^{\varkappa,n} := \mathfrak{F}_{+}^{\varkappa,n} \cap \mathfrak{F}_{-}^{\varkappa,n} , \quad \mathfrak{F}_{\cup}^{\varkappa,n} := \langle \mathfrak{F}_{+}^{\varkappa,n} \cup \mathfrak{F}_{-}^{\varkappa,n} \rangle .$$

Here $\langle \mathfrak{M} \rangle$ means the linear hull of $\mathfrak{M}$. One trivially has

$$\text{(1.3.5)} \qquad \mathfrak{F}_{\cap}^{\varkappa,n} = \{0\} \text{ for } n+\varkappa+1 > 0, \quad \mathfrak{F}_{\cup}^{\varkappa,n} = \mathfrak{F} \quad \text{for } n < 0 .$$

$\mathfrak{F}$ or subspaces of $\mathfrak{F}$ of the kind just described will be contained in the Hilbert spaces of all irreducible unitary representations, to be constructed in the following, as dense subspaces.

A transitive operation of SU(1,1) on the boundary of the unit circle is defined by

$$\text{(1.3.6)} \qquad SU(1,1) \ni A: \omega \longrightarrow \omega\bar{A} := \frac{\omega A_{11}+A_{21}}{\omega A_{12}+A_{22}}$$

A quasi-invariant measure on the boundary of the unit circle is $d\omega/\omega$ with the Radon-Nikodym derivative

$$\text{(1.3.7)} \qquad d\omega\bar{A}/\omega\bar{A} = |\omega A_{12}+A_{22}|^{-2} \, d\omega/\omega \quad .$$

A continuous linear representation of SU(1,1) on $\mathfrak{F}$, for $\varkappa \in \{0,1\}$ and $\ell \in \mathbb{C}$, is defined by

$$SU(1,1) \ni A \longrightarrow V^{\varkappa,\ell}(A):$$

(1.3.8)

$$(V^{\varkappa,\ell}(A)f)(\omega) = |\omega A_{12}+A_{22}|^{-2\ell-\varkappa-2} \left(\frac{\omega A_{12}+A_{22}}{|\omega A_{12}+A_{22}|} \right)^{\varkappa} f(\omega\bar{A}) \; .$$

According to [14], Chapter VII,2, for noninteger ℓ this representation is irreducible in the sense, that $\mathfrak{F}$ has no proper subspace invariant under $V^{\varkappa,\ell}$. For integer ℓ the subspaces $\mathfrak{F}_{\pm}^{\varkappa,\ell}$, $\mathfrak{F}_{\cap}^{\varkappa,\ell}$ and $\mathfrak{F}_{\cup}^{\varkappa,\ell}$ are invariant under $V^{\varkappa,\ell}$. For <u>$\ell+\varkappa+1>0$</u> the restrictions

$$\text{(1.3.9)} \qquad V_{\pm}^{\varkappa,\ell} := V^{\varkappa,\ell} \,|\, \mathfrak{F}_{\pm}^{\varkappa,\ell} , \; \ell+\varkappa+1 > 0 \text{ , integer}$$

prove to be irreducible subrepresentations. Since in this case $\mathcal{F}_{\cup}^{\varkappa,\ell}$ (except for $\varkappa=1$, $\ell=-1$) is a proper invariant subspace, $V^{\varkappa,\ell}$ in a natural way induces a representation on the quotient space $\mathcal{F}/\mathcal{F}_{\cup}^{\varkappa,\ell}$ which proves to be irreducible. Because of the isomorphism

$$\mathcal{F}/\mathcal{F}_{\cup}^{\varkappa,\ell} \cong \mathcal{F}_{\cap}^{\varkappa,-\ell-\varkappa-1}, \quad \ell+\varkappa+1>0, \text{ integer} \tag{1.3.10}$$

we can consider this also on the representants taken from $\mathcal{F}_{\cap}^{\varkappa,-\ell-\varkappa-1}$. So besides $V_{\pm}^{\varkappa,\ell}$ we get a third irreducible representation

$$V_{\cap}^{\varkappa,\ell} := P_{\cap}^{\varkappa,-\ell-\varkappa-1} V^{\varkappa,\ell} P_{\cap}^{\varkappa,-\ell-\varkappa-1} \big| \mathcal{F}_{\cap}^{\varkappa,-\ell-\varkappa-1}, \quad \ell+\varkappa+1>0, \text{ integer}, \tag{1.3.11}$$

i.e. the restriction to $\mathcal{F}_{\cap}^{\varkappa,-\ell-\varkappa-1}$ of the representation $V^{\varkappa,\ell}$, reduced by the projectors $P_{\cap}^{\varkappa,-\ell-\varkappa-1}$ to $\mathcal{F}_{\cap}^{\varkappa,-\ell-\varkappa-1}$. For $\underline{\ell<0}$ we have a dual situation. Here $\mathcal{F}_{\cap}^{\varkappa,\ell}$ (except for $\varkappa=1$, $\ell=-1$) proves to be a proper invariant subspace of $\mathcal{F}$, and the restriction

$$V_{\cap}^{\varkappa,\ell} := V^{\varkappa,\ell} \big| \mathcal{F}_{\cap}^{\varkappa,\ell}, \quad \ell<0, \text{ integer} \tag{1.3.12}$$

is an irreducible representation of SU(1,1). Since in this case the $\mathcal{F}_{\pm}^{\varkappa,\ell}$ contain $\mathcal{F}_{\cap}^{\varkappa,\ell}$ as a proper subspace, the restrictions of $V^{\varkappa,\ell}$ to $\mathcal{F}_{\pm}^{\varkappa,\ell}$ are not irreducible. But the representations induced in a natural way by $V^{\varkappa,\ell}$ on the quotient spaces $\mathcal{F}_{\pm}^{\varkappa,\ell}/\mathcal{F}_{\cap}^{\varkappa,\ell}$ prove to be irreducible. Because of the isomorphisms

$$\mathcal{F}_{\pm}^{\varkappa,\ell}/\mathcal{F}_{\cap}^{\varkappa,\ell} \cong \mathcal{F}_{\pm}^{\varkappa,-\ell-\varkappa-1} \cong \mathcal{F}/\mathcal{F}_{\mp}^{\varkappa,\ell}, \quad \ell<0, \text{ integer} \tag{1.3.13}$$

we can consider these representations on the representants chosen from $\mathcal{F}_{\pm}^{\varkappa,-\ell-\varkappa-1}$. So besides $V_{\cap}^{\varkappa,\ell}$ we get the irreducible representations

$$V_{\pm}^{\varkappa,\ell} := P_{\pm}^{\varkappa,-\ell-\varkappa-1} V^{\varkappa,\ell} P_{\pm}^{\varkappa,-\ell-\varkappa-1} \big| \mathcal{F}_{\pm}^{\varkappa,-\ell-\varkappa-1}, \quad \ell<0, \text{ integer}, \tag{1.3.14}$$

the restrictions to $\mathcal{F}_{\pm}^{\varkappa,-\ell-\varkappa-1}$ of the representation $V^{\varkappa,\ell}$, reduced by the projectors $P_{\pm}^{\varkappa,-\ell-\varkappa-1}$ to $\mathcal{F}_{\pm}^{\varkappa,-\ell-\varkappa-1}$.

For integer ℓ we therefore have a decomposition

$$\mathcal{F} = \begin{cases} \mathcal{F}_{+}^{\varkappa,\ell} \oplus \mathcal{F}_{-}^{\varkappa,\ell} \oplus \mathcal{F}_{\cap}^{\varkappa,-\ell-\varkappa-1} & \text{for } \ell+\varkappa+1>0, \text{integer}, \\ \mathcal{F}_{+}^{\varkappa,-\ell-\varkappa-1} \oplus \mathcal{F}_{-}^{\varkappa,-\ell-\varkappa-1} \oplus \mathcal{F}_{\cap}^{\varkappa,\ell} & \text{for } \ell<0, \text{ integer} \end{cases} \tag{1.3.15}$$

of $\mathfrak{F}$ into the direct sum of three subspaces, on each of which an irreducible representation of SU(1,1) is given. Here the pairs of representations belonging to ℓ and $-\ell-\varkappa-1$ are defined on the same subspaces. It should be noted, however, that $V^{\varkappa,\ell}$ is not decomposable and therefore in no way decomposes into the direct sum of the subrepresentations $V_{\pm}^{\varkappa,\ell}$, $V_{\cap}^{\varkappa,\ell}$. For the case $\varkappa = 1$, $\ell = -1$ in which the cases $\ell+\varkappa+1>0$ and $\ell<0$ overlap we still have to specify the above statements. Here $\mathfrak{F}_{\cap}^{1,-1} = \{0\}$ and $\mathfrak{F}$ decays into the direct sum $\mathfrak{F}_{+}^{1,-1} \oplus \mathfrak{F}_{-}^{1,-1}$. We then have only the two irreducible representations $V_{\pm}^{1,-1}$, and we even get $V^{1,-1} = V_{+}^{1,-1} \oplus V_{-}^{1,-1}$.

The following assertion concerning the existence of bilinear functionals, invariant under a pair of representations, which was proven in [14], Chapter VII,3 for the group SL(2,$\mathbb{R}$) which is isomorphic with SU(1,1), can be regarded as a decisive theorem for the representation theory of SU(1,1):

A bilinear functional on $\mathfrak{F}$ of the form

$$\langle f \mid Bg\rangle := (2\pi i)^{-2} \oint\oint \frac{d\omega'}{\omega'}\frac{d\omega}{\omega} f(\omega')^{*} B(\omega',\omega)\, g(\omega) \tag{1.3.16}$$

with a generalized integral kernel $B(\omega',\omega)$ which is invariant under a pair $V^{\varkappa,\ell}$, $V^{\varkappa',\ell'}$ of representations, i.e. which fulfils the condition

$$\langle V^{\varkappa',\ell'}(A)f \mid BV^{\varkappa,\ell}(A)g\rangle = \langle f \mid Bg\rangle \quad , \ A \in SU(1,1) \tag{1.3.17}$$

exists if and only if $\varkappa' = \varkappa$ and $\ell' = -\ell^{*}-\varkappa-1$ or $\ell' = \ell^{*}$. In the case $\underline{\ell' = -\ell^{*}-\varkappa-1}$ such a functional is defined by

$$B(\omega',\omega) = 2\pi\,\delta(-i \ln\omega'/\omega) \quad , \tag{1.3.18}$$

and any other one is proportional to it except for $\varkappa = 1$, $\ell = -1$. In the case $\underline{\ell' = \ell^{*}, \ell \text{ noninteger}}$, such a functional, which also is unique up to a factor, is defined by

$$B(\omega',\omega) = T^{\varkappa,\ell}(\omega'/\omega) \tag{1.3.19}$$

with

$$T^{\varkappa,\ell}(\omega) := \frac{\pi\,|1-\omega|^{2\ell}(1-\omega)^{\varkappa}}{\Gamma(2\ell+\varkappa+1)\sin\pi(\ell+\varkappa+1)} \tag{1.3.20}$$

For $\underline{\ell' = \ell < 0, \text{ integer}}$, there exist on $\mathcal{F}$, up to a constant factor, exactly the two invariant functionals

$$B = T_{+}^{\varkappa,\ell} \quad \text{and} \quad B = T_{-}^{\varkappa,\ell} , \tag{1.3.21}$$

where the $T_{\pm}^{\varkappa,\ell}$ are defined by the "spectral representation"

$$T_{+}^{\varkappa,\ell}\, e_\nu = \begin{cases} \dfrac{(\nu-\ell-\varkappa-1)!}{(\nu+\ell)!}\, e_\nu , & \nu \geq -\ell , \\ 0 , & \text{otherwise}, \end{cases}$$

(1.3.22)

$$T_{-}^{\varkappa,\ell}\, e_\nu = \begin{cases} (-)^{\varkappa}\dfrac{(-\nu-\ell-1)!}{(-\nu+\ell+\varkappa)!}\, e_\nu , & \nu \leq \ell+\varkappa , \\ 0 , & \text{otherwise}, \quad \ell<0, \text{ integer}. \end{cases}$$

Therefore $T_{\pm}^{\varkappa,\ell}$ is degenerate on $\mathcal{F}_{\mp}^{\varkappa,\ell}$; both functionals are degenerate on $\mathcal{F}_{\cap}^{\varkappa,\ell}$. On this subspace, which is nontrivial for $\ell+\varkappa+1 \leq 0$, there exists, uniquely up to a factor, the invariant functional defined by the "spectral representation"

$$\hat{T}_{\cap}^{\varkappa,\ell}\, e_\nu = (-)^{\nu+\ell+1}\,(\nu-\ell-\varkappa-1)!\,(-\nu-\ell-1)!\, e_\nu ,$$

(1.3.23)

$$\ell+\varkappa+1 \leq \nu \leq -\ell-1 , \quad \ell+\varkappa+1 \leq 0 , \quad \text{integer}.$$

For $\underline{\ell' = \ell \geq 0, \text{ integer}}$, exists, uniquely up to a factor, the invariant functional

$$B = T_{\cap}^{\varkappa,\ell} \tag{1.3.24}$$

which is defined on $\mathcal{F}$ by

$$T_{\cap}^{\varkappa,\ell}\, e_\nu = \begin{cases} (-)^{\nu-\ell-\varkappa}\dfrac{1}{(\ell+\varkappa-\nu)!\,(\ell+\nu)!}\, e_\nu , & -\ell \leq \nu \leq \ell+\varkappa , \\ 0 , & \text{otherwise}, \quad \ell \geq 0, \text{integer}. \end{cases} \tag{1.3.25}$$

$T_{\cap}^{\varkappa,\ell}$ is degenerate on $\mathcal{F}_{+}^{\varkappa,\ell}$ and $\mathcal{F}_{-}^{\varkappa,\ell}$. On both of these subspaces a functional, invariant there and unique up to a factor, can be introduced by the definition

$$(1.3.26)\qquad \begin{aligned} \hat{T}_{+}^{\varkappa,\ell}\, e_\nu &= \frac{(\nu-\ell-\varkappa-1)!}{(\nu+\ell)!}\, e_\nu\,, \quad \nu \geq \ell+\varkappa+1,\\ \hat{T}_{-}^{\varkappa,\ell}\, e_\nu &= (-)^{\varkappa}\frac{(-\nu-\ell-1)!}{(-\nu+\ell+\varkappa)!}\, e_\nu\,, \quad \nu \leq -\ell-1\,, \quad \ell \geq 0, \text{ integer}\,. \end{aligned}$$

Before we use the above theorem for the representation theory of SU(1,1), we discuss some properties of the functional $T^{\varkappa,\ell}$. For noninteger ℓ the functional $T^{\varkappa,\ell}$, as defined by (1.3.20), is also diagonal in the basis $\{e_\nu\}$:

$$(1.3.27)\qquad T^{\varkappa,\ell}\, e_\nu = \frac{\Gamma(\nu-\ell-\varkappa)}{\Gamma(\nu+\ell+1)}\, e_\nu = (-)^{\varkappa}\frac{\Gamma(-\nu-\ell)}{\Gamma(-\nu+\ell+\varkappa+1)}\, e_\nu\,.$$

From this at once follows, that $T^{\varkappa,-\ell-\varkappa-1}$ is the inverse operator of $T^{\varkappa,\ell}$:

$$(1.3.28)\qquad T^{\varkappa,\ell}\, T^{\varkappa,-\ell-\varkappa-1} = \mathbb{1}_{\mathcal{F}} = T^{\varkappa,-\ell-\varkappa-1}\, T^{\varkappa,\ell}\,.$$

Using the "hat" symbol generally for the restrictions, i.e. defining still

$$(1.3.29)\qquad \begin{aligned} \hat{T}_{\pm}^{\varkappa,\ell} &:= T_{\pm}^{\varkappa,\ell}\big|\mathcal{F}_{\pm}^{\varkappa,-\ell-\varkappa-1} && \text{for } \ell<0, \text{ integer},\\ \hat{T}_{\cap}^{\varkappa,\ell} &:= T_{\cap}^{\varkappa,\ell}\big|\mathcal{F}_{\cap}^{\varkappa,-\ell-\varkappa-1} && \text{for } \ell\geq 0, \text{ integer}, \end{aligned}$$

one easily sees, that the operators $\hat{T}_{\pm}^{\varkappa,\ell}$ and $\hat{T}_{\cap}^{\varkappa,\ell}$ can be interpreted as "limits" of the operators $T^{\varkappa,z}$ for complex z:

$$(1.3.30)\qquad \begin{aligned} \hat{T}_{\pm}^{\varkappa,\ell} &= \begin{cases} \lim\limits_{z\to\ell} T^{\varkappa,z}\big|\mathcal{F}_{\pm}^{\varkappa,-\ell-\varkappa-1}, & \ell<0, \text{ integer},\\ \lim\limits_{z\to\ell} T^{\varkappa,z}\big|\mathcal{F}_{\pm}^{\varkappa,\ell}, & \ell\geq 0, \text{ integer}, \end{cases}\\ \hat{T}_{\cap}^{\varkappa,\ell} &= \begin{cases} \lim\limits_{z\to\ell} \frac{d}{dz}\, T^{\varkappa,z}\big|\mathcal{F}_{\cap}^{\varkappa,\ell}, & \ell<0, \text{ integer},\\ -\operatorname*{Res}\limits_{z=\ell} T^{\varkappa,z}\big|\mathcal{F}_{\cap}^{\varkappa,-\ell-\varkappa-1}, & \ell\geq 0, \text{ integer}. \end{cases} \end{aligned}$$

Also the restricted operators belonging to ℓ and $-\ell-\varkappa-1$ are reciprocals of each other:

$$(1.3.31)\qquad \begin{aligned} \hat{T}_{\pm}^{\varkappa,\ell}\hat{T}_{\pm}^{\varkappa,-\ell-\varkappa-1} = \hat{T}_{\pm}^{\varkappa,-\ell-\varkappa-1}\hat{T}_{\pm}^{\varkappa,\ell} &= \begin{cases} \mathbb{1}_{\mathcal{F}_{\pm}^{\varkappa,-\ell-\varkappa-1}}, & \ell<0, \text{ integer},\\ \mathbb{1}_{\mathcal{F}_{\pm}^{\varkappa,\ell}}, & \ell\geq 0, \text{ integer}, \end{cases}\\ \hat{T}_{\cap}^{\varkappa,\ell}\hat{T}_{\cap}^{\varkappa,-\ell-\varkappa-1} = \hat{T}_{\cap}^{\varkappa,-\ell-\varkappa-1}\hat{T}_{\cap}^{\varkappa,\ell} &= \begin{cases} \mathbb{1}_{\mathcal{F}_{\cap}^{\varkappa,\ell}}, & \ell<0, \text{ integer},\\ \mathbb{1}_{\mathcal{F}_{\cap}^{\varkappa,-\ell-\varkappa-1}}, & \ell\geq 0, \text{ integer}. \end{cases} \end{aligned}$$

As a first consequence of the theorem cited above we show that exactly the representations belonging to ℓ and $-\ell-\varkappa-1$, respectively, are equivalent. Because of the rule

$$\langle V^{\varkappa,\ell}(A)f \mid Bg\rangle = \langle f \mid V^{\varkappa,-\ell^*-\varkappa-1}(A^{-1})Bg\rangle \tag{1.3.32}$$

holding for every B in (1.3.16) the invariant functionals existing in the case $\ell' = \ell^*$ define intertwining operators between the corresponding representations, while B is proportional to the unit operator for $\ell' = -\ell^*-\varkappa-1$. For noninteger ℓ we therefore have

$$T^{\varkappa,\ell}\, V^{\varkappa,\ell} = V^{\varkappa,-\ell-\varkappa-1}\, T^{\varkappa,\ell} , \tag{1.3.33}$$

and for the subrepresentations belonging to integer ℓ we get

$$\begin{aligned} \hat{T}^{\varkappa,\ell}_{\pm}\, V^{\varkappa,\ell}_{\pm} &= V^{\varkappa,-\ell-\varkappa-1}_{\pm}\, \hat{T}^{\varkappa,\ell}_{\pm} , \\ \hat{T}^{\varkappa,\ell}_{\cap}\, V^{\varkappa,\ell}_{\cap} &= V^{\varkappa,-\ell-\varkappa-1}_{\cap}\, \hat{T}^{\varkappa,\ell}_{\cap} . \end{aligned} \tag{1.3.34}$$

A second consequence from the above theorem is the Schur irreducibility of all representations $V^{\varkappa,\ell}$. For in the case $\ell' = -\ell^*-\varkappa-1$ the δ-functional (1.3.18) is the invariant functional, uniquely defined up to a factor, and therefore because of the rule (1.3.32) the multiples of the unit operator are the only operators which commute with the representation $V^{\varkappa,\ell}$. An exception is the case $\varkappa = 1$, $\ell = -1$ in which the operators $\pm T^{1,-1}_{\pm}$ are the projectors upon the subspaces $\mathcal{F}^{1,-1}_{\pm}$ and commute with $V^{1,-1}$ because of (1.3.34). This corresponds to the above remark on $V^{1,-1}$ being the direct sum $V^{1,-1}_{+} \oplus V^{1,-1}_{-}$.

Finally the theorem may be used for the examination under which conditions the representations constructed above on $\mathcal{F}$ and on subspaces of $\mathcal{F}$ can be extended to unitary Hilbert space representations. For the unitary representations obtained in this way we will sketch, later on, a proof of their irreducibility.

The scalar product of a Hilbert space, of which $\mathcal{F}$ is a dense subspace and in which $U^{\varkappa,\ell}_{SU(1,1)}$ is a unitary representation the restriction to $\mathcal{F}$ of which is identical with $V^{\varkappa,\ell}$, is a special bilinear functional of the type described in the theorem with $\varkappa' = \varkappa$, $\ell' = \ell$. It is symmetrical and positive definite:

(1.3.35) $$\langle f|Bg\rangle = \langle g|Bf\rangle^* \ , \quad \langle f|Bf\rangle > 0 \quad \text{for } f \neq 0 \ .$$

According to the theorem it exists only in the case $\mathrm{Re}\,\ell = -(1+\varkappa)/2$ or $\mathrm{Im}\,\ell = 0$. In the first case B is the δ-functional which obviously is symmetrical and positive definite. In the second case, for noninteger ℓ, the functional $T^{\varkappa,\ell}$ is symmetrical because it has real eigenvalues according to (1.3.27). However, it is positive definite only for $\varkappa = 0$, $-1 < \ell < 0$. For the ratio $(\nu-\ell-\varkappa)/(\nu+\ell+1)$ of the eigenvalues belonging to $e_{\nu+1}$ and e_ν is positive for all ν only for these values of $\varkappa$ and ℓ, and for $\varkappa = 0$, $\ell = -1/2$ $T^{\varkappa,\ell}$ is the unit operator. For integer ℓ we consider the functionals $\hat{T}_\pm^{\varkappa,\ell}$ and $\hat{T}_\cap^{\varkappa,\ell}$ defined on subspaces. The three functionals are symmetrical because of their real eigenvalues. On $\mathcal{F}_\cap^{\varkappa,\ell}$ for $\ell < 0$ and on $\mathcal{F}_\cap^{\varkappa,-\ell-\varkappa-1}$ for $\ell+\varkappa+1 > 0$ a positive definite functional exists only in the case $\varkappa = 0$, $\ell = -1$ and $\varkappa = 0$, $\ell = 0$, respectively, because of the alternating signs of the eigenvalues. On the other hand on $\mathcal{F}_\pm^{\varkappa,\ell}$ for $\ell+\varkappa+1 > 0$ and on $\mathcal{F}_\pm^{\varkappa,-\ell-\varkappa-1}$ for $\ell < 0$ the operators $(\pm)^\varkappa\hat{T}_\pm^{\varkappa,\ell}$ for $\varkappa \in \{0,1\}$ and all integer ℓ are positive definite. In the cases just singled out the invariant, symmetrical and positive definite functionals on $\mathcal{F}$ or subspaces of $\mathcal{F}$ can be introduced as scalar products. The completions of the linear spaces relative to the corresponding scalar product are Hilbert spaces $\mathfrak{h}_{SU(1,1)}^{\varkappa,\ell}$ on which the representations $V^{\varkappa,\ell}$ can be extended to unitary representations $U_{SU(1,1)}^{\varkappa,\ell}$. We therefore get the following series of unitary representations of SU(1,1):

1. <u>The principal series:</u> It consists of the representations

(1.3.36)
$$(U_{SU(1,1)}^{\varkappa,\ell}(A)f)(\omega) = \sqrt{\frac{d\omega\bar{A}}{\omega\bar{A}}\Big/\frac{d\omega}{\omega}}\,|\omega A_{12}+A_{22}|^{-2\ell-\varkappa-1}\left(\frac{\omega A_{12}+A_{22}}{|\omega A_{12}+A_{22}|}\right)^{\varkappa} f(\omega\bar{A}),$$
$$\varkappa = 0,1;\quad \ell = -(1+\varkappa)/2+i\xi\ ,\ -\infty < \xi < \infty\ ,$$

on the Hilbert space $\mathfrak{h}_{SU(1,1)}^{\varkappa,\ell} \equiv \mathfrak{h}$ of square-integrable functions on the boundary of the unit circle with scalar product

(1.3.37) $$\langle f|g\rangle_{SU(1,1)}^{\varkappa,\ell} := \frac{1}{2\pi i}\oint\frac{d\omega}{\omega}\, f(\omega)^*\, g(\omega)\ .$$

2. <u>The supplementary series:</u> It consists of the representations

(1.3.38)
$$(U_{SU(1,1)}^{0,\ell}(A)f)(\omega) = \frac{d\omega\bar{A}}{\omega\bar{A}}\Big/\frac{d\omega}{\omega}\,|\omega A_{12}+A_{22}|^{-2\ell} f(\omega\bar{A})\ ,$$
$$\varkappa = 0;\ -1 < \ell < 0,$$

on the Hilbert spaces $\mathfrak{h}^{0,\ell}_{SU(1,1)}$ with the scalar products

$$\langle f|g\rangle^{0,\ell}_{SU(1,1)} := \langle f|T^{0,\ell}g\rangle = \sum_{\nu=-\infty}^{+\infty} \frac{\Gamma(\nu-\ell)}{\Gamma(\nu+\ell+1)} f^*_\nu g_\nu \,. \tag{1.3.39}$$

3. <u>The discrete series:</u> It consists of the representations

$$(U^{\varkappa,\ell,\pm}_{SU(1,1)}(A)f)(\omega) = \frac{1}{2\pi i}\oint \frac{d\omega'}{\omega'} P^{\varkappa,\ell_0}_{\pm}\left(\frac{\omega}{\omega'}\right)|\omega' A_{12}+A_{22}|^{-2\ell-\varkappa-1} \times \left(\frac{\omega' A_{12}+A_{22}}{|\omega' A_{12}+A_{22}|}\right)^{\varkappa} f(\omega'\bar{A}) \,, \tag{1.3.40}$$

$$\varkappa = 0,1; \quad \ell = 0,\pm1,\pm2,\dots\,; \quad \ell_0 := \max(\ell,-\ell-\varkappa-1) \,,$$

where $P^{\varkappa,\ell_0}_{\pm}(\omega/\omega')$ is the kernel belonging to the projector onto $\mathfrak{F}^{\varkappa,\ell_0}_{\pm}$, $\ell_0 := \max(\ell,-\ell-\varkappa-1)$, on the Hilbert spaces $\mathfrak{h}^{\varkappa,\ell,\pm}_{SU(1,1)}$ which originate from the completion of the just mentioned spaces relative to the scalar products

$$\langle f|g\rangle^{\varkappa,\ell,\pm}_{SU(1,1)} := \langle f|(\pm)^{\varkappa}\hat{T}^{\varkappa,\ell}_{\pm}g\rangle = \sum_{\nu=\max(-\ell,\ell+\varkappa+1)}^{+\infty} \frac{(\nu-\ell-\varkappa-1)!}{(\nu+\ell)!} f^*_{\pm(\nu+\varkappa/2)+\varkappa/2}\, g_{\pm(\nu+\frac{\varkappa}{2})+\frac{\varkappa}{2}} \,. \tag{1.3.41}$$

For $\ell+\varkappa+1>0$ the projector in (1.3.40) can be omitted.

4. <u>The unit representations:</u> These are the representations

$$(U^{0,\ell,\cap}_{SU(1,1)}(A)f)(\omega) = \frac{1}{2\pi i}\oint \frac{d\omega'}{\omega'} P^{0,0}_{\cap}(\omega/\omega')\,|\omega' A_{12}+A_{22}|^{-2\ell-2} f(\omega'\bar{A}), \qquad \varkappa = 0; \quad \ell = 0,-1, \tag{1.3.42}$$

on the one-dimensional Hilbert space $\mathfrak{h}^{0,\ell,\cap}_{SU(1,1)} \equiv \mathfrak{F}^{0,0}_{\cap} \equiv \mathbb{C}$. The projector $P^{0,0}_{\cap}$ onto $\mathfrak{F}^{0,0}_{\cap}$ can be omitted for $\ell = -1$.

Principal and supplementary series intersect in the point $\varkappa = 0$, $\ell = -1/2$. The corresponding representations of both series are identical. Principal and discrete series intersect in the point $\varkappa = 1, \ell = -1$. The representation $U^{1,-1}_{SU(1,1)}$ of the principal series decomposes into the direct sum of the representations $U^{1,-1,+}_{SU(1,1)}$ and $U^{1,-1,-}_{SU(1,1)}$ of the discrete series. The equivalence relations (1.3.33) and (1,3,34) may be extended for the unitary series to unitary equivalence relations. Hence the representations belonging to ℓ and $-\ell-\varkappa-1$ are unitary equivalent to each other. The equivalence classes of irreducible unitary representations of SU(1,1) therefore are defined already by one half of the above series. For the parameter ρ, introduced in Section 1.1 to characterize the equivalence classes of irreducible unitary representations of the little groups, we may fix the following convention:

$$(1.3.43)\quad \rho \longrightarrow (\varkappa,\ell,\eta): \begin{cases} \varkappa = 0,1;\ \ell = -(1+\varkappa)/2+i\xi,\ \xi \geq 0,\ \eta = 0; & \text{principal series;} \\ \varkappa = 0;\ -1/2 < \ell < 0;\ \eta = 0; & \text{supplementary series;} \\ \varkappa = 0,1;\ \ell = 0,1,2,\ldots;\ \eta = \pm; & \text{discrete series;} \\ \varkappa = 0;\ \ell = -1;\ \eta = \cap; & \text{unit representation.} \end{cases}$$

Here we have introduced $\eta = 0$ as a dead parameter for the principal and supplementary series for the sake of a unified notation $\mathfrak{h}^{\varkappa,\ell,\eta}_{SU(1,1)}$ and $U^{\varkappa,\ell,\eta}_{SU(1,1)}$ for the representation spaces and representations. In the following we will discuss only the representations belonging to the parameters given in (1.3.43). For these the projectors in (1.3.40) and (1.3.42) can be omitted. The unit representation, however, will not be considered henceforth.

Up to the special case $\varkappa = 1$, $\ell = -1$ of the principal series, all representations of the above series are irreducible. The proof rests upon the extension of Schurs lemma to unitary representations on Hilbert spaces. According to it a representation is irreducible if and only if the multiples of the unit operator are the only bounded operators which commute with it. Let C be a bounded operator that commutes with the representation $U^{\varkappa,\ell,\eta}_{SU(1,1)}$. From the commutability of C with the restriction of $U^{\varkappa,\ell,\eta}_{SU(1,1)}$ to the diagonal subgroup $H_1 := \{A \in SU(1,1) : A_{12} = A^*_{21} = 0\}$ follows that C must be diagonal in the basis $\{e_\nu\}$. For the eigenvalues c_ν of C we then get the relation

$(c_{\nu'}-c_\nu) \langle e_{\nu'}| U^{\varkappa,\ell,\eta}_{SU(1,1)}(A)\, e_\nu \rangle^{\varkappa,\ell,\eta}_{SU(1,1)} = 0$ for every $A \in SU(1,1)$. We will calculate the matrix elements explicitly in Section 2.3 and can anticipate that indeed in the range corresponding to the representation $U^{\varkappa,\ell,\eta}_{SU(1,1)}$ of the indices ν',ν they are different from zero generally. Therefore C must be a multiple of the unit operator.

The representations which belong to the parameters quoted in (1.3.43) are representants of all equivalence classes of irreducible unitary representations of SU(1,1), that is, every irreducible unitary representation of SU(1,1) is equivalent to one of the above. An explicit proof of this assertion can be found e.g. in TAKAHASHI [16].

For later use we shall present still another realization of the representations of the discrete series. To this purpose we continue analytically the elements of the dense subset $\mathcal{F}_{+}^{\varkappa,\ell}$ of $\mathfrak{h}_{SU(1,1)}^{\varkappa,\ell,+}$ into the interior of the unit circle $\{z \in \mathbb{C} : |z| < 1\}$ by adjoining to $f = \sum_{\ell+\varkappa+1}^{\infty} f_\nu e_\nu \in \mathcal{F}_{+}^{\varkappa,\ell}$ the function $f' = \sum_{\ell+\varkappa+1}^{\infty} f_\nu e'_\nu$, $e'_\nu(z) := z^\nu$. By completion relative to the scalar product

$$\langle f' | g' \rangle'^{\varkappa,\ell,+}_{SU(1,1)} := \int_{|z|<1} d\mu^{\varkappa,\ell}(z)\, f'(z)^* g'(z) ,$$

$$(1.3.44) \quad \langle e'_{\nu'} | e'_\nu \rangle'^{\varkappa,\ell,+}_{SU(1,1)} = \delta_{\nu'\nu} \frac{(\nu-\ell-\varkappa-1)!}{(\nu+\ell)!} = \langle e_{\nu'} | e_\nu \rangle^{\varkappa,\ell,+}_{SU(1,1)} ,$$

$$d\mu^{\varkappa,\ell}(z) := \frac{dx\, dy\, (1-|z|^2)^{2\ell+\varkappa}}{\pi (2\ell+\varkappa)!\, |z|^{2\ell+2\varkappa+2}} , \quad z = x + iy ,$$

we get a Hilbert space $\mathfrak{h}'^{\varkappa,\ell,+}_{SU(1,1)}$ of functions holomorphic in the interior of the unit circle with a zero of order $\geq \ell+\varkappa+1$ in the origin. A Hilbert space isomorphism between $\mathfrak{h}^{\varkappa,\ell,+}_{SU(1,1)}$ and $\mathfrak{h}'^{\varkappa,\ell,+}_{SU(1,1)}$ is given by

$$(1.3.45) \quad \begin{aligned} \mathfrak{h}^{\varkappa,\ell,+}_{SU(1,1)} \ni f = \sum_{\ell+\varkappa+1}^{\infty} f_\nu e_\nu &\longrightarrow f' = \sum_{\ell+\varkappa+1}^{\infty} f_\nu e'_\nu \in \mathfrak{h}'^{\varkappa,\ell,+}_{SU(1,1)} , \\ \mathfrak{h}'^{\varkappa,\ell,+}_{SU(1,1)} \ni f' &\longrightarrow f : f(\omega) = \lim_{z\to\omega} f'(z) \in \mathfrak{h}^{\varkappa,\ell,+}_{SU(1,1)} . \end{aligned}$$

Now the transitive operation $z \longrightarrow z\bar{A} = (zA_{11}+A_{21})/(zA_{12}+A_{22})$ of SU(1,1) on the interior of the unit circle can be considered as an analytical continuation of the corresponding operation $\omega \longrightarrow \omega\bar{A}$ on the boundary of the unit circle. Because ℓ is an integer, the representation (1.3.40) can be written in the form $(U^{\varkappa,\ell,+}_{SU(1,1)}(A)f)(\omega) = (\omega A_{12}+A_{22})^{-\ell-1}(\omega^{-1}A_{21}+A_{11})^{-\ell-\varkappa-1}\, f(\omega\bar{A})$, which apparently can be continued analytically into the interior of the unit circle to a representation

$$(1.3.46) \quad (U'^{\varkappa,\ell,+}_{SU(1,1)}(A)f')(z) = (zA_{12}+A_{22})^{-\ell-1}(z^{-1}A_{21}+A_{11})^{-\ell-\varkappa-1} f'(z\bar{A})$$

on $\mathfrak{h}'^{\varkappa,\ell,+}_{SU(1,1)}$. The latter, because of (1.3.45) and (1.3.44), is unitary equivalent to $U^{\varkappa,\ell,+}_{SU(1,1)}$. In a similar way we could realize $U^{\varkappa,\ell,-}_{SU(1,1)}$ by analytical continuation into the exterior of the unit circle. But we choose another way in transforming first $U^{\varkappa,\ell,-}_{SU(1,1)}$ by the unitary mapping

(1.3.47) $$L_\varkappa:\ \mathfrak{h}^{\varkappa,\ell,-}_{SU(1,1)} \longrightarrow \mathfrak{h}^{\varkappa,\ell,+}_{SU(1,1)}\ ,\quad (L_\varkappa f)(\omega) := \omega^{\varkappa} f(\omega^{-1})$$

into the representation

(1.3.48) $$L_\varkappa U^{\varkappa,\ell,-}_{SU(1,1)} L_\varkappa^{-1} = U^{\varkappa,\ell,+\,*}_{SU(1,1)}\ ,\quad U^{\varkappa,\ell,+\,*}_{SU(1,1)}(A) := U^{\varkappa,\ell,+}_{SU(1,1)}(A^*)$$

and then continuing this into the interior of the unit circle to the representation

(1.3.49) $$U'^{\varkappa,\ell,-}_{SU(1,1)} = U'^{\varkappa,\ell,+\,*}_{SU(1,1)}\ ,\quad U'^{\varkappa,\ell,+\,*}_{SU(1,1)}(A) := U'^{\varkappa,\ell,+}_{SU(1,1)}(A^*)\ .$$

We also point out the antiunitary equivalence of the representations $U'^{\varkappa,\ell,+}_{SU(1,1)}$ and $U'^{\varkappa,\ell,-}_{SU(1,1)}$, for with the antiunitary operator

(1.3.50) $$K'_\varkappa:\ \mathfrak{h}'^{\varkappa,\ell,+}_{SU(1,1)} \longrightarrow \mathfrak{h}'^{\varkappa,\ell,+}_{SU(1,1)},\quad (K'_\varkappa f)(z) := f(z^*)^*\ ,\quad K'^{\,2}_\varkappa = \mathbb{1}_{\mathfrak{h}'^{\varkappa,\ell,+}_{SU(1,1)}}$$

we get the relation

(1.3.51) $$K'_\varkappa\, U'^{\varkappa,\ell,+}_{SU(1,1)}(A)\, K'_\varkappa = U'^{\varkappa,\ell,+}_{SU(1,1)}(A^*) = U'^{\varkappa,\ell,-}_{SU(1,1)}(A)\ .$$

Finally we remark, that the reproducing kernel corresponding to the unit operator in $\mathfrak{h}'^{\varkappa,\ell,+}_{SU(1,1)}$ is given by

(1.3.52) $$K^{\varkappa,\ell,+}(z',z) := \sum_{\ell+\varkappa+1}^{\infty} \frac{(\nu+\ell)!}{(\nu-\ell-\varkappa-1)!} e'_\nu(z') e'_\nu(z)^* = \frac{(2\ell+\varkappa+1)!\,(z'z^*)^{\ell+\varkappa+1}}{(1-z'z^*)^{2\ell+\varkappa+2}}\ .$$

1.4 The Irreducible Unitary Representations of E(2)

For the elements of E(2) we choose the notation

(1.4.1) $$E(2) \ni (e^{i\varphi/2}, z) = A\ ,\quad e^{i\varphi/2} := A_{11} = A^*_{22}\ ,\quad z = A_{11}A_{12}\ .$$

This simultaneously means a parametrization of E(2) where

(1.4.2) $$0 \leq \varphi < 4\pi\ ,\qquad z \in \mathbb{C}\ .$$

The group law is given by

(1.4.3) $$(e^{i\varphi/2}, z)\,(e^{i\varphi'/2}, z') = (e^{i(\varphi+\varphi')/2}, z + e^{i\varphi} z')$$

so that E(2) is isomorphic to the semidirect product $SO(2) ⓢ \mathbb{C}$ of the plane rotation group SO(2) and the additive group $\mathbb{C}$ of complex numbers with the operation d: $SO(2) \longrightarrow \mathrm{Aut}\,\mathbb{C}$, $d_\varphi(z) := e^{i\varphi}z$ of SO(2) on $\mathbb{C}$. Therefore all irreducible unitary representations of E(2) can be obtained by the same method as sketched in Section 1.1 for the Poincaré group itself.

I. The characters of the abelian invariant subgroup $(1,\mathbb{C}) \subset E(2)$ are

$$(1.4.4) \qquad (1,z) \longrightarrow \chi^{\zeta}(z) = e^{i\,\mathrm{Re}(\zeta^* z)}, \qquad \zeta \in \mathbb{C},$$

i.e. the character group is the additive group $\mathbb{C}$ itself.

II. The orbits of SO(2) on the character group $\mathbb{C}$, because of $\chi^{\zeta}(e^{-i\varphi}z) = \chi^{\zeta e^{i\varphi}}(z)$, are the circles $\omega(\zeta)$ around the origin through ζ: $\omega(\zeta) = SO(2)\zeta$.

III. The partition of $\mathbb{C}$ into orbits is characterized by the following set of representants:

$$(1.4.5) \quad \mathbb{C} = \bigcup_{\rho\in\omega} \omega(\rho), \quad \omega = \omega^+ \cup \omega^0, \quad \omega^+ = \{\rho : \rho > 0\}, \quad \omega^0 = \{0\}.$$

IV. The little group $G(\rho)$ belonging to $\rho \in \omega$ is calculated from the condition $\chi^{\rho}(e^{-i\varphi}z) = \chi^{\rho}(z)$ for every $e^{i\varphi/2} \in G(\rho)$ and $z \in \mathbb{C}$. One finds

$$(1.4.6) \quad \begin{aligned} \rho\in\omega^+ &: G(\rho) = \mathbb{Z}_2 := \{1,-1\}, \\ \rho\in\omega^0 &: G(\rho) = G(0) = SO(2). \end{aligned}$$

V. The left coset from $SO(2)/G(\rho)$ belonging to $\zeta \in \omega(\rho)$, $\rho > 0$, is represented by $\exp(i\varphi_\zeta/2)$, $\varphi_\zeta := \mathrm{arc}\,\zeta$, $0 \le \varphi_\zeta < 2\pi$, for obviously $d_{\varphi_\zeta}(\rho) = \zeta$. The case $\rho = 0$ is trivial.

VI. The irreducible unitary representations of $\mathbb{Z}_2$ have the form

$$(1.4.7) \qquad \mathbb{Z}_2 \ni \varepsilon \longrightarrow U^{\varkappa}_{\mathbb{Z}_2}(\varepsilon) = \varepsilon^{\varkappa}, \quad \varkappa = 0,1.$$

$G(0) \equiv SO(2)$ has the irreducible unitary representations

$$(1.4.8) \quad SO(2) \ni e^{i\varphi/2} \longrightarrow U^{\varkappa,\mu}_{SO(2)}(e^{i\varphi/2}) = e^{i(\mu+\varkappa/2)\varphi}, \quad \varkappa = 0,1;\ \mu = 0,\pm1,\ldots$$

VII. For the group $\check{G}(\rho) := G(\rho) ⓢ \mathbb{C} \subset E(2)$ we therefore get the representations

$$\rho\in\omega^+ : \check{G}(\rho) \ni (\varepsilon, z) \longrightarrow U^{\rho,\varkappa}_{\check{G}(\rho)}(\varepsilon,z) = \varepsilon^{\varkappa}\chi^{\rho}(z) \quad , \varkappa = 0,1;$$

(1.4.9)

$$\rho\in\omega^0 : \check{G}(0) \ni (e^{i\varphi/2}, z) \longrightarrow U^{0,\varkappa,\mu}_{\check{G}(0)}(e^{i\varphi/2},z) = e^{i(\mu+\varkappa/2)\varphi}, \varkappa = 0,1; \quad \mu = 0,\pm 1,\ldots$$

VIII. In the case $\rho > 0$ we set up the Hilbert space

(1.4.10) $$\mathfrak{h}^{\rho,\varkappa}_{E(2)} : = \oplus \int_0^{2\pi} \sqrt{\frac{d\varphi}{2\pi}}\, \mathfrak{h}^{\varkappa}_{G(\rho)}(\rho e^{i\varphi}) ,$$

the direct integral of the one-dimensional Hilbert spaces $\mathfrak{h}^{\varkappa}_{G(\rho)}(\zeta) \equiv$ $\equiv \mathbb{C}$. It has the scalar product

(1.4.11) $$\langle f | g \rangle^{\rho,\varkappa}_{E(2)} : = \int_0^{2\pi} \frac{d\varphi}{2\pi} f(\rho e^{i\varphi})^* g(\rho e^{i\varphi}) .$$

In the case $\rho = 0$ there is no direct integral since $\omega(0)$ consists of one point only.

IX. The induced representations of E(2) are for $\rho > 0$:

$$(\bar{U}^{\rho,\varkappa}_{E(2)}(e^{i\varphi/2},z)f)(\zeta) = \varepsilon(\zeta,\varphi)^{\varkappa}\, e^{i\,\mathrm{Re}(\zeta^* z)}\, f(e^{-i\varphi}\zeta),$$

(1.4.12) $$\varkappa = 0,1; \quad |\zeta| = \rho ;$$

$$\varepsilon(\zeta,\varphi) : = e^{i(\varphi+\varphi_{e^{-i\varphi}\zeta}-\varphi_\zeta)/2} = \begin{cases} +1 \text{ for } 0\le\varphi\le\varphi_\zeta \text{ and } 2\pi+\varphi_\zeta<\varphi<4\pi , \\ -1 \text{ for } \varphi_\zeta<\varphi\le\varphi_\zeta+2\pi , \end{cases}$$

and for $\rho = 0$:

(1.4.13) $$U^{0,\varkappa,\mu}_{E(2)}(e^{i\varphi/2},z) = e^{i(\mu+\varkappa/2)\varphi} , \ \varkappa = 0,1; \ \mu = 0,\pm 1,\pm 2, \ldots$$

For the description of massless particles one usually takes the one-dimensional representations of the last kind. In these the abelian invariant subgroup $(1,\mathbb{C}) \subset E(2)$ is represented trivially.

The somewhat unconvenient form of the representation $\bar{U}^{\rho,\varkappa}_{E(2)}$ in (1.4.12) can be smoothed with the aid of the unitary transformation

(1.4.14) $$\tau^{\varkappa} : f \longrightarrow \tau^{\varkappa} f , \ (\tau^{\varkappa} f)(\zeta) : = e^{i\varkappa\varphi_\zeta/2} f(\zeta)$$

in the Hilbert space $\mathfrak{h}^{\rho,\varkappa}_{E(2)}$. The representation

(1.4.15) $$U^{\rho,\varkappa}_{E(2)} : = \tau^{\varkappa}\, \bar{U}^{\rho,\varkappa}_{E(2)}\, \tau^{\varkappa\dagger}$$

which is equivalent to $\overline{U}^{\rho,\varkappa}_{E(2)}$ has the explicit form

$$(U^{\rho,\varkappa}_{E(2)}(e^{i\varphi/2},z)f)(\zeta) = e^{i\varkappa\varphi/2 + i\,\mathrm{Re}(\zeta^* z)}\, f(e^{-i\varphi}\zeta) \; ,$$

(1.4.16)

$$\varkappa = 0,1; \quad |\zeta| = \rho \; .$$

1.5 The Irreducible Unitary Representations of SL(2,ℂ)

The irreducible unitary representations of SL(2,ℂ) were set up first by GELFAND and NEUMARK [17]. Referring to the detailed description of the representation theory of SL(2,ℂ) in the monograph [18] by NEUMARK we may be brief here.

A transitive operation of the group SL(2,ℂ) in the closed complex plane $\overline{\mathbb{C}}$ is defined by

$$(1.5.1) \quad SL(2,\mathbb{C}) \ni A: \; z \longrightarrow z\overline{A}: = (zA_{11}+A_{21})/(zA_{12}+A_{22}) \; , \quad z \in \overline{\mathbb{C}} \; .$$

The quasi-invariant measure on $\overline{\mathbb{C}}$ up to equivalence is given by

$$(1.5.2) \qquad d\mu(z): = \pi^{-1}(1+|z|^2)^{-2}\, dx\, dy \; , \quad z = x + iy$$

with the Radon-Nikodym derivative

$$(1.5.3) \qquad d\mu(z\overline{A})/d\mu(z) = |zA_{12}+A_{22}|^{-4} \left(\frac{1+|z|^2}{1+|z\overline{A}|^2}\right)^2 .$$

The irreducible unitary representations of SL(2,ℂ) can be realized in spaces of complex valued functions on $\overline{\mathbb{C}}$ (cf. NEUMARK [18]). Apart from the unit representation they have the form

$$(U^{\varkappa,\lambda,\mu}_{SL(2,\mathbb{C})}(A)f)(z) = \sqrt{d\mu(z\overline{A})/d\mu(z)}^{\,1+i\lambda} \left(\frac{zA_{12}+A_{22}}{|zA_{12}+A_{22}|}\right)^{2\mu+\varkappa} f(z\overline{A}) \; ,$$

(1.5.4)

$$\varkappa \in \{0,1\}; \; \mu \in \{0,\pm 1,\pm 2, \ldots\}; \; \lambda \in \mathbb{R} \cup \{-it: \; 0 < t < 1\} \; .$$

One distinguishes between the following series:

1. The principal series: It consists of the representations $U^{\varkappa,\lambda,\mu}_{SL(2,\mathbb{C})}$ with $\lambda \in \mathbb{R}$ on the Hilbert space $\mathfrak{h}^{\varkappa,\lambda,\mu}_{SL(2,\mathbb{C})} \equiv \mathfrak{h}$ with the scalar product

$$(1.5.5) \qquad \langle f|g\rangle^{\varkappa,\lambda,\mu}_{SL(2,\mathbb{C})}: = \int_{\overline{\mathbb{C}}} d\mu(z)\, f(z)^* g(z) \quad ;$$

2. The supplementary series: It consists of the representations $U^{0,\lambda,0}_{SL(2,\mathbb{C})}$ with $\lambda = -it$, $0 < t < 1$ on the Hilbert spaces $\mathfrak{h}^{0,\lambda,0}_{SL(2,\mathbb{C})}$ with the scalar products

$$\langle f|g\rangle^{0,\lambda,0}_{SL(2,\mathbb{C})} := \int\limits_{\overline{\mathbb{C}}\times\overline{\mathbb{C}}}\int d\mu(z')d\mu(z)\ f(z')^* K_\lambda(z',z)\ g(z)\ ,$$

(1.5.6)

$$\Gamma(i\lambda)\ K_\lambda(z',z) := \left(\frac{|z'-z|^2}{(1+|z'|^2)(1+|z|^2)}\right)^{-1+i\lambda} .$$

Two representations $U^{\varkappa',\lambda',\mu'}_{SL(2,\mathbb{C})}$ and $U^{\varkappa,\lambda,\mu}_{SL(2,\mathbb{C})}$ are equivalent if and only if either $(\varkappa',\lambda',\mu') = (\varkappa,\lambda,\mu)$ or $(\varkappa',\lambda',\mu') = (\varkappa,-\lambda,-\mu-\varkappa)$.

The well known finite-dimensional, nonunitary irreducible representations of $SL(2,\mathbb{C})$ can be realized on the linear spaces

$$L^{j_1,j_2} := \left\{f:\overline{\mathbb{C}}\longrightarrow\mathbb{C}:\ f(z) = (1+|z|^2)^{-j_1-j_2}\sum_{n=0}^{2j_1}\sum_{n=0}^{2j_2} a_{n'n}\, z^{n'} z^{*n},\ a_{n'n}\in\mathbb{C}\right\},$$

(1.5.7)

$$j_1,\ j_2 = 0,1/2,1,3/2,\ \ldots$$

They are of the form given in (1.5.4) with $\lambda = i(1+j_1+j_2)$, $\mu+\varkappa/2 = j_1-j_2$.

2. Matrix Elements of the Irreducible Unitary Representations of the Little Groups and Expansion Theorems for Square-Integrable Functions on Coset Spaces of the Little Groups

In this Chapter we derive expansion theorems as needed in Chapter 3 for square-integrable functions on certain coset spaces of the little groups. The coset spaces are defined by the subgroups

$$\begin{aligned} H_1 &:= SU(2)\cap SU(1,1) = SU(2)\cap E(2) = E(2)\cap SU(1,1)\ , \\ H_2 &:= SU(1,1)\cap SL(2,\mathbb{R})\ . \end{aligned} \tag{2.1}$$

We first solve the reduction problem for the restriction of the little group representation U_G^ρ, $G \in \{G(\mathring{p}): \mathring{p}\in\Omega\ ,\ H\subset G(\mathring{p})\}$ to the subgroup H, $H \in \{H_1, H_2\}$, where ρ is a parameter running through the set $\hat{G}$ of all equivalence classes of irreducible unitary representations of G. Since all groups discussed here are of type I (cf. MACKEY [10]), the reduction problem is uniquely solvable, i.e. there exists a unitary equivalence transformation $\breve{A}$ which carries $U_G^\rho|H$ over to a direct integral of irreducible unitary representations χ^σ of H:

$$\breve{A}\,\mathfrak{h}_G^\rho = \bigoplus_{\hat{H}}\int \sqrt{d\tilde{\nu}_\rho(\sigma)}\ \mathfrak{h}_G^{\rho,\sigma},\quad \breve{A}\ U_G^\rho|H\ \breve{A}^{-1} = \bigoplus_{\hat{H}}\int d\tilde{\nu}_\rho(\sigma)\,\chi^\sigma\ . \tag{2.2}$$

Here σ is a parameter running through the set $\hat{H}$ of equivalence classes of irreducible unitary representations of H, $\tilde{\nu}_\rho$ a measure on H defined by the decomposition, and $\mathfrak{h}_G^{\rho,\sigma}$ is a component of the representation space $\breve{A}\,\mathfrak{h}_G^\rho$ on which a multiple of χ^σ operates. Let

$$\{\psi_{\tau\sigma}^\rho:\ \tau = 1,2,\ \dots\ ,\ \dim\mathfrak{h}_G^{\rho,\sigma}\}\ ,\quad \langle\psi_{\tau'\sigma}^\rho|\psi_{\tau\sigma}^\rho\rangle_{\mathfrak{h}_G^{\rho,\sigma}} = \delta_{\tau'\tau}\ , \tag{2.3}$$

be an orhtonormal basis in $\mathfrak{h}_G^{\rho,\sigma}$. Then for $f \in \mathfrak{h}_G^\rho$ we have the expansion

$$\breve{A}\,f = \bigoplus_{\hat{H}}\int\sqrt{d\tilde{\nu}_\rho(\sigma)}\sum_\tau \psi_{\tau\sigma}^\rho\, f_{\tau\sigma}\ . \tag{2.4}$$

As matrix elements of U_G^ρ relative to a basis belonging to the subgroup H we denote the generalized integral kernels $U_G^\rho(A)_{\tau'\sigma',\tau\sigma}$ defined by

$$(U_G^\rho(A)f)_{\tau'\sigma'} = \int_{\hat{H}} d\tilde{\nu}_\rho(\sigma)\sum_\tau U_G^\rho(A)_{\tau'\sigma',\tau\sigma} f_{\tau\sigma}\ . \tag{2.5}$$

(For the compact group H_1 the direct integrals are reduced to direct sums. The matrix elements exist in the usual sense.) We calculate them in Sections 2.2 - 2.5. In a certain generalized sense they are complete orthonormal bases in the Hilbert spaces of square-integrable functions on the little groups. This is shown in Section 2.6. In Section 2.7 we use these bases for a generalized Fourier analysis of square-integrable functions on the coset spaces G/H.

The matrix elements of the SU(2)-representations relative to a H_1-basis are the elements of the representation matrices of the rotation group, well known from quantum mechanics. The matrix elements of the SU(1,1)-representations relative to a H_1-basis were calculated first by BARGMANN [13]. A solution of the reduction problem for the restrictions of the SU(1,1)-representations to H_2 was given by MUKUNDA [19] by means of the Lie algebra. His result confirms BARGMANNs [13] assertion concerning the multiplicity of the spectra of the generators of one parameter subgroups of SU(1,1). In Section 2.5 we give a global solution of the reduction problem for $U^{\rho}_{SU(1,1)}|H_2$ and calculate the corresponding generalized matrix elements. Some of the latter have been calculated previously by VILENKIN [20] and used for the derivation of formulas for the hypergeometric functions ${}_2F_1$. The Plancherel formula for square-integrable functions on SU(2) is contained in the famous Peter-Weyl theorem for compact Lie groups. For the noncompact group SU(1,1) it first was proved explicitly by HARISH-CHANDRA [21]. The representations needed for the expansion were found previously by BARGMANN [13]. We give in Section 2.6 another proof, using the analytical properties of the matrix elements which are of special interest for the S-matrix theory of elementary particles. It permits to handle the SU(2)- and SU(1,1)-representations simultaneously. The use of matrix elements of the SU(2)-representations for the expansion of square-integrable functions on the sphere was mentioned in GELFAND, MINLOS and SHAPIRO [22]. We know of no other literature on expansion theorems for square-integrable functions on the homogeneous spaces G/H as set up in Section 2.7.

2.1 The Irreducible Unitary Representations of the Groups $H_1 := SU(2)\cap SU(1,1)$ and $H_2 := SU(1,1)\cap SL(2,\mathbb{R})$

The intersection of the groups $G(e_{(0)}) = SU(2)$, $G(e_{(0)}+e_{(3)}) = E(2)$ and $G(e_{(3)}) = SU(1,1)$ is the group

$$(2.1.1)\qquad H_1 := \left\{C(\varphi) := \begin{pmatrix} e^{i\varphi/2} & 0 \\ 0 & e^{-i\varphi/2}\end{pmatrix} : 0 \le \varphi < 4\pi\right\}.$$

Obviously it is isomorphic to the plane euclidean rotation group SO(2). The irreducible unitary representations are

(2.1.2)
$$C(\varphi) \longrightarrow U^{\varkappa,\mu}(C(\varphi)) \equiv \chi^{\varkappa,\mu}(\varphi) = e^{i(\mu+\varkappa/2)\varphi} ,$$
$$\varkappa = 0,1; \ \mu = 0,\pm 1,\pm 2, \ \ldots \ .$$

They obey the orthogonality and completeness relations

(2.1.3)
$$\int_0^{4\pi} \frac{d\varphi}{4\pi} \chi^{\varkappa',\mu'}(-\varphi)\, \chi^{\varkappa,\mu}(\varphi) = \delta_{\varkappa'\varkappa}\, \delta_{\mu'\mu}$$

and

(2.1.4)
$$\sum_{\varkappa=0,1} \sum_{\mu=-\infty}^{+\infty} \chi^{\varkappa,\mu}(\varphi)\, \chi^{\varkappa,\mu}(-\varphi') = 4\pi\, \delta(\varphi-\varphi') \ .$$

The groups $G(e_{(2)}) = SL(2,\mathbb{R})$ and $G(e_{(3)}) = SU(1,1)$ have the intersection group

(2.1.5)
$$H_2 := \left\{ D(\varepsilon,\xi) := \varepsilon \begin{pmatrix} \cosh\xi/2 & \sinh\xi/2 \\ \sinh\xi/2 & \cosh\xi/2 \end{pmatrix} : \varepsilon = \pm 1, \ -\infty < \xi < \infty \right\}.$$

Because of the group law

(2.1.6)
$$D(\varepsilon',\xi')\, D(\varepsilon,\xi) = D(\varepsilon'\varepsilon, \xi'+\xi)$$

H_2 is isomorphic to the direct product $\mathbb{Z}_2 \otimes \mathbb{R}$ of the cyclic group of order two, $\mathbb{Z}_2$, and the additive group of real numbers, $\mathbb{R}$. The irreducible unitary representations of H_2 are

(2.1.7)
$$D(\varepsilon,\xi) \longrightarrow U^{\varkappa,\lambda}(D(\varepsilon,\xi)) \equiv \chi^{\varkappa,\lambda}(\varepsilon,\xi) = \varepsilon^{\varkappa}\, e^{i\lambda\xi} ,$$
$$\varkappa = 0,1; \ -\infty < \lambda < \infty \ .$$

They obey the orthogonality and completeness relations

(2.1.8)
$$\frac{1}{2} \sum_{\varepsilon=\pm} \int_{-\infty}^{+\infty} \frac{d\xi}{2\pi} \chi^{\varkappa',\lambda'}(\varepsilon,-\xi)\, \chi^{\varkappa,\lambda}(\varepsilon,\xi) = \delta_{\varkappa'\varkappa}\delta(\lambda'-\lambda)$$

and

(2.1.9)
$$\frac{1}{2} \sum_{\varkappa=0,1} \int_{-\infty}^{+\infty} \frac{d\lambda}{2\pi} \chi^{\varkappa,\lambda}(\varepsilon,\xi)\, \chi^{\varkappa,\lambda}(\varepsilon',-\xi') = \delta_{\varepsilon'\varepsilon}\, \delta(\lambda'-\lambda) \ .$$

2.2 Matrix Elements of the Irreducible Unitary Representations of SU(2) Relative to an H_1-Basis

An orthonormal basis of $\mathfrak{h}^{\varkappa,\ell}_{SU(2)}$ is

$$\left\{\psi^{\varkappa,\ell}_{\mu}: -\ell-\varkappa \leq \mu \leq \ell,\ \psi^{\varkappa,\ell}_{\mu}(z_1,z_2):= \frac{z_1^{\ell+\varkappa+\mu} z_2^{\ell-\mu}}{[(\ell+\varkappa+\mu)!(\ell-\mu)!]^{1/2}}\right\}, \tag{2.2.1}$$
$$\langle\psi^{\varkappa,\ell}_{\mu'}|\psi^{\varkappa,\ell}_{\mu}\rangle^{\varkappa,\ell}_{SU(2)} = \delta_{\mu'\mu}.$$

With (1.2.2) and (1.2.3) we obviously get

$$U^{\varkappa,\ell}_{SU(2)}(C(\varphi))\,\psi^{\varkappa,\ell}_{\mu} = \chi^{\varkappa,\mu}(\varphi)\ \psi^{\varkappa,\ell}_{\mu}, \tag{2.2.2}$$

i.e. (2.2.1) defines a basis belonging to H_1. An irreducible unitary representation $\chi^{\varkappa',\mu}$ of H_1 is found in $U^{\varkappa,\ell}_{SU(2)}$ exactly once if $\varkappa' = \varkappa$ and $-\ell-\varkappa\leq\mu\leq\ell$, otherwise not.

The matrix elements of $U^{\varkappa,\ell}_{SU(2)}$ relative to the above basis are obtained from

$$\begin{aligned}(U^{\varkappa,\ell}_{SU(2)}(A)\psi^{\varkappa,\ell}_{\mu})(z_1,z_2) &= \frac{(z_1A_{11}+z_2A_{21})^{\ell+\varkappa+\mu}(z_1A_{12}+z_2A_{22})^{\ell-\mu}}{[(\ell+\varkappa+\mu)!(\ell-\mu)!]^{1/2}} = \\ &= \sum_{\mu'=-\ell-\varkappa}^{\ell}\psi^{\varkappa,\ell}_{\mu'}(z_1,z_2)\,U^{\varkappa,\ell}_{SU(2)}(A)_{\mu'\mu}\ ; \\ U^{\varkappa,\ell}_{SU(2)}(A)_{\mu'\mu} &:= \langle\psi^{\varkappa,\ell}_{\mu'}|U^{\varkappa,\ell}_{SU(2)}(A)\psi^{\varkappa,\ell}_{\mu}\rangle^{\varkappa,\ell}_{SU(2)},\end{aligned} \tag{2.2.3}$$

with the aid of the binomial expansion. The result is

$$\begin{aligned}U^{\varkappa,\ell}_{SU(2)}(A)_{\mu'\mu} &= (A_{11}/|A_{11}|)^{\mu'+\mu+\varkappa}(A_{12}/|A_{12}|)^{\mu'-\mu}\,\tilde{u}'^{\varkappa,\ell}_{\mu'\mu}(|A_{12}|^2), \\ \tilde{u}'^{\varkappa,\ell}_{\mu'\mu}(x) &:= +\sqrt{\frac{\Gamma(1+\ell+\mu'+\varkappa)\Gamma(1+\ell-\mu)}{\Gamma(1+\ell+\mu+\varkappa)\Gamma(1+\ell-\mu')}}\ x^{(\mu'-\mu)/2}(1-x)^{(\mu'+\mu+\varkappa)/2}\ \times \\ &\times \frac{1}{\Gamma(\mu'-\mu+1)}\ {}_2F_1(\mu'-\ell,\mu'+\ell+\varkappa+1;\mu'-\mu+1;x), \quad 0\leq x\leq 1\ .\end{aligned} \tag{2.2.4}$$

Here the hypergeometric function ${}_2F_1$ reduces to a polynomial because $\mu'-\ell$ is a nonpositive integer. With regard to the analytical continuation of the matrix elements $U^{\varkappa,\ell}_{SU(2)}(A)_{\mu'\mu}$, discussed in Section 2.6, we write (2.2.4) in the somewhat modified form

$$U^{\varkappa,\ell}_{SU(2)}(A)_{\mu'\mu} = (A_{11}/A_{22})^{(\mu'+\mu+\varkappa)/2}(A_{12}/A_{21})^{(\mu'-\mu)/2}\tilde{u}^{\varkappa,\ell}_{\mu'\mu}(-A_{12}A_{21}), \tag{2.2.5}$$

where $\tilde{u}^{\varkappa,\ell}_{\mu'\mu}$ is the single-valued analytical function

$$\tilde{u}^{\varkappa,\ell}_{\mu'\mu}(z) := +\sqrt{\frac{\Gamma(1+\ell+\mu'+\varkappa)\Gamma(1+\ell-\mu)}{\Gamma(1+\ell+\mu+\varkappa)\Gamma(1+\ell-\mu')}}\,(-z)^{(\mu'-\mu)/2}(1-z)^{(\mu'+\mu+\varkappa)/2}\times \tag{2.2.6}$$
$$\times\frac{1}{\Gamma(\mu'-\mu+1)}\,{}_2F_1(\mu'-\ell,\mu'+\ell+\varkappa+1;\mu'-\mu+1;z)$$

which is defined in the complex z-plane cut from 0 to $+\infty$ along the real axis with real values on the negative real axis. For the function

$$\tilde{u}'^{\varkappa,\ell}_{\mu'\mu}(z) := e^{i\pi\,\mathrm{sign}(\mathrm{Im}\,z)(\mu'-\mu)/2}\,\tilde{u}^{\varkappa,\ell}_{\mu'\mu}(z) \tag{2.2.7}$$

we have apparently $\tilde{u}'^{\varkappa,\ell}_{\mu'\mu}(x\pm io) = \tilde{u}'^{\varkappa,\ell}_{\mu'\mu}(x)$, $0\leq x\leq 1$, and therefore (2.2.5) for $A_{21} = A^*_{12}\,e^{\pm i\pi}$ coincides with (2.2.4).

To derive symmetry relations for the matrix elements (2.2.5) we consider the linear operator $T_\varkappa$ and the antilinear operator $K_\varkappa$ on $\mathfrak{h}^{\varkappa,\ell}_{SU(2)}$ defined by

$$(T_\varkappa f)(z_1,z_2) = f(-z_2,z_1)\;, \qquad (K_\varkappa f)(z_1,z_2) = f(-z_2^*,z_1^*)^*\;, \tag{2.2.8}$$
$$T_\varkappa\,\psi^{\varkappa,\ell}_{\mu} = (-)^{\ell+\mu+\varkappa}\,\psi^{\varkappa,\ell}_{-\mu-\varkappa} = K_\varkappa\,\psi^{\varkappa,\ell}_{\mu}\;.$$

They fulfil the relation

$$T^2_\varkappa = (-)^\varkappa\,\mathbf{1}_{\mathfrak{h}^{\varkappa,\ell}_{SU(2)}} = K^2_\varkappa\;. \tag{2.2.9}$$

While $T_\varkappa$ connects the representation $U^{\varkappa,\ell}_{SU(2)}$ with its contragredient, $K_\varkappa$ commutes with $U^{\varkappa,\ell}_{SU(2)}$:

$$T_\varkappa\,U^{\varkappa,\ell}_{SU(2)}(A)\,T^{-1}_\varkappa = U^{\varkappa,\ell}_{SU(2)}(A^{-1T}),\quad K_\varkappa\,U^{\varkappa,\ell}_{SU(2)}(A)\,K^{-1}_\varkappa = U^{\varkappa,\ell}_{SU(2)}(A). \tag{2.2.10}$$

From this follow for the matrix elements (2.2.5) the symmetry relations

$$U^{\varkappa,\ell}_{SU(2)}(A^{-1T})_{\mu'\mu} = (-)^{\mu'-\mu}\,U^{\varkappa,\ell}_{SU(2)}(A)_{-\mu'-\varkappa,-\mu-\varkappa} = U^{\varkappa,\ell}_{SU(2)}(A^\dagger)_{\mu\mu'}, \tag{2.2.11}$$
$$\tilde{u}^{\varkappa,\ell}_{\mu'\mu}(z) = \tilde{u}^{\varkappa,\ell}_{-\mu'-\varkappa,-\mu-\varkappa}(z) = \tilde{u}^{\varkappa,\ell}_{\mu\mu'}(z)\;.$$

Of course these relations could have been obtained from a table of formulas for the hypergeometric functions as well. The procedure applied here is an example of a strongly generalizable method to derive conversely, with the aid of the representation theory of groups, formulas for special functions.

2.3 Matrix Elements of the Irreducible Unitary Representations of SU(1,1) Relative to an H_1-Basis

According to (1.3.8) on the basis $\{e_\nu\}$ of $\mathcal{F}$ we have

$$(2.3.1) \qquad (V^{\varkappa,\ell}(C(\varphi))e_\nu)(\omega) = \chi^{\varkappa,\nu-\varkappa}(\varphi)\, e_\nu(\omega) \,.$$

This apparently means for the irreducible unitary representations $U^{\varkappa,\ell,\eta}_{SU(1,1)}$, obtained from the $V^{\varkappa,\ell}$, that the irreducible unitary representation $\chi^{\varkappa',\mu}$ of H_1 appears exactly once in $U^{\varkappa,\ell,\eta}_{SU(1,1)}$ if $\varkappa' = \varkappa$ and if $e_{\mu+\varkappa}$ is contained in the representation space $\mathfrak{H}^{\varkappa,\ell,\eta}_{SU(1,1)}$. So $\{e_\nu\}$ is a basis belonging to H_1. Because of (1.3.36) - (1.3.41), however, the basis elements e_ν have different normalizations in the different Hilbert spaces. If we disregard the unit representation, we can write these in the form

$$(2.3.2) \qquad \langle e_{\nu'}|e_\nu \rangle^{\varkappa,\ell,\eta}_{SU(1,1)} = \delta_{\nu'\nu}\left|\frac{\Gamma(\nu-\ell-\varkappa)}{\Gamma(\nu+\ell+1)}\right| \,.$$

We therefore use in the following the orthonormal bases of the Hilbert spaces $\mathfrak{H}^{\varkappa,\ell,\eta}_{SU(1,1)}$ defined by

$$(2.3.3) \quad \left\{\psi^{\varkappa,\ell,\eta}_\mu = N^{\varkappa,\ell}_\mu\, e_{\mu+\varkappa}\colon \mu = \begin{cases} 0,\pm1,\pm2,\,\ldots & \text{for } \eta = 0 \,, \\ +1,\,+2,\,\ldots & \text{for } \eta = + \,; \\ -\,-\,-1,\,-\,-\,-2,\,\ldots & \text{for } \eta = - \,; \end{cases}\right.$$

$$\left|N^{\varkappa,\ell}_\mu\right|^2 = \left|\frac{\Gamma(\mu+\ell+\varkappa+1)}{\Gamma(\mu-\ell)}\right|\right\} \,, \quad \langle\psi^{\varkappa,\ell,\eta}_{\mu'}|\psi^{\varkappa,\ell,\eta}_\mu\rangle^{\varkappa,\ell,\eta}_{SU(1,1)} = \delta_{\mu'\mu}.$$

The normalization factor $N^{\varkappa,\ell}_\mu$ is to be understood as a restriction to the ℓ-values of the unitary series of the function

$$(2.3.4) \qquad N^{\varkappa,\ell}_\mu\colon = \left[\frac{\Gamma(\mu+\ell+\varkappa+1)}{\Gamma(\mu-\ell)}\right]^{1/2}$$

on the complex ℓ-plane. The function $\Gamma(\mu+\ell+\varkappa+1)/\Gamma(\mu-\ell)$ has zeros at the points μ, $\mu+1$, ... and poles at the points $-\mu-\varkappa-1$, $-\mu-\varkappa-2$, ... For $\mu \le -1$ both chains overlap, and the zeros and poles lying between μ and $-\mu-\varkappa-1$ compensate. On the real axis the denominator is positive for $\ell < \mu$, the numerator is positive for $\ell > -\mu-\varkappa-1$, while they have alternating signs between the zeros and the poles. At the point $\ell = -(1+\varkappa)/2$ the function $\Gamma(\mu+\ell+\varkappa+1)/\Gamma(\mu-\ell)$ assumes the value +1 for $\mu \geq 0$ and $(-)^{\varkappa}$ for $-\mu-\varkappa \geq 0$. If we make the convention

$$N_{\mu}^{\varkappa,\ell}\Big|_{\ell=-(1+\varkappa)/2\pm io} = \begin{cases} 1 & \text{for } \mu \geq 0, \\ e^{\pm i\pi(\mu+\varkappa/2)} & \text{for } \mu < 0, \end{cases} \tag{2.3.5}$$

the function $N_{\mu}^{\varkappa,\ell}$ is single-valued in the complex ℓ-plane cut along the intervals of the real axis where $\Gamma(\mu+\ell+\varkappa+1)/\Gamma(\mu-\ell)$ is negative. With the convention (2.3.5) for numerator and denominator separately we have $[\Gamma(\mu+\ell+\varkappa+1)]^{1/2} > 0$ for $\ell > -\mu-\varkappa-1$ and $[\Gamma(\mu-\ell)]^{1/2} > 0$ for $\ell < \mu$. The following symmetry relations hold for $N_{\mu}^{\varkappa,\ell}$:

$$\begin{gathered} (N_{\mu}^{\varkappa,\ell})^* = N_{\mu}^{\varkappa,\ell^*}\ , \quad N_{\mu}^{\varkappa,-\ell-\varkappa-1}\, N_{\mu}^{\varkappa,\ell} = 1\ , \\ N_{-\mu-\varkappa}^{\varkappa,\ell} = N_{\mu}^{\varkappa,\ell}\, e^{-i\pi \operatorname{sign}(\operatorname{Im}\ell)(\mu+\varkappa/2)}\ . \end{gathered} \tag{2.3.6}$$

The matrix elements of the unitary series of SU(1,1)-representations relative to the bases (2.3.3) have the common form

$$\begin{aligned} U_{SU(1,1)}^{\varkappa,\ell,\eta}(A)_{\mu'\mu} :&= \langle \psi_{\mu'}^{\varkappa,\ell,\eta} | U_{SU(1,1)}^{\varkappa,\ell,\eta}(A)\, \psi_{\mu}^{\varkappa,\ell,\eta} \rangle_{SU(1,1)}^{\varkappa,\ell,\eta} = \\ &= N_{\mu}^{\varkappa,\ell}/N_{\mu'}^{\varkappa,\ell}\; V_{\mu'\mu}^{\varkappa,\ell}(A)\ , \\ V_{\mu'\mu}^{\varkappa,\ell}(A) :&= |N_{\mu'}^{\varkappa,\ell}|^2 \langle e_{\mu'+\varkappa} | V^{\varkappa,\ell}(A)\, e_{\mu+\varkappa} \rangle_{SU(1,1)}^{\varkappa,\ell,\eta} = \\ &= \frac{1}{2\pi i}\oint \frac{d\omega}{\omega}\, e_{\mu'+\varkappa}(\omega)^* |\omega A_{12}+A_{22}|^{-2\ell-\varkappa-2} \left(\frac{\omega A_{12}+A_{22}}{|\omega A_{12}+A_{22}|}\right)^{\varkappa} e_{\mu+\varkappa}(\omega \bar{A})\ . \end{aligned} \tag{2.3.7}$$

Here the matrix elements $V_{\mu'\mu}^{\varkappa,\ell}(A)$ exist for any $\ell \in \mathbb{C}$, for because of the inequality

$$0 < |A_{22}| - |A_{12}| \leq |\omega A_{12}+A_{22}| \leq |A_{22}| + |A_{12}| \tag{2.3.8}$$

we have the estimate

$$|V_{\mu'\mu}^{\varkappa,\ell}(A)| \leq \frac{1}{2\pi i}\oint \frac{d\omega}{\omega} |\omega A_{12}+A_{22}|^{-2\operatorname{Re}\ell-\varkappa-2} \leq (|A_{22}| + |A_{12}|)^{|2\operatorname{Re}(\ell+\varkappa+1)|}. \tag{2.3.9}$$

Therefore we calculate $V_{\mu'\mu}^{\varkappa,\ell}(A)$ for any complex ℓ. To this purpose we continue analytically $|\omega A_{12}+A_{22}|^{-2\ell-2\varkappa-2}$ on both sides of the unit circle $\{\omega \in \mathbb{C} : |\omega| = 1\}$ into the annular domain

$\{\omega \in \mathbb{C}: |A_{12}/A_{11}| < |\omega| < |A_{22}/A_{12}|\}$. We write it in the form $(\omega A_{12}+A_{22})^{-\ell-\varkappa-1}(\omega^{-1}A_{21}+A_{11})^{-\ell-\varkappa-1}$ and define both factors by $\exp[-(\ell+\varkappa+1)\ln(\omega A_{12}+A_{22})]$ and $\exp[-(\ell+\varkappa+1)\ln(\omega^{-1}A_{21}+A_{11})]$, respectively, with the principal value of the logarithm. Then $V^{\varkappa,\ell}_{\mu'\mu}(A)$ takes the form

$$(2.3.10) \quad V^{\varkappa,\ell}_{\mu'\mu}(A) = \frac{1}{2\pi i}\oint \frac{d\omega}{\omega}\,\omega^{\mu-\mu'}(\omega A_{12}+A_{22})^{-\ell-\varkappa-1-\mu}(\omega^{-1}A_{21}+A_{11})^{-\ell-1+\mu}.$$

The integrand is single-valued in the ω-plane cut from 0 to $-A_{21}/A_{11}$ and from $-A_{22}/A_{12}$ to ∞, if on the straight line $\{\omega = rA_{21}/A_{11}: r>0\}$ we fix the phases in the following way: $\operatorname{arc}(\omega A_{12}+A_{22}) = \operatorname{arc} A_{22}$, $\operatorname{arc}(\omega^{-1}A_{21}+A_{11}) = \operatorname{arc} A_{11}$. The integration path encloses the first cut. By substituting

$$(2.3.11) \qquad t = -A_{12}{}^{-1}(\omega A_{11}+A_{21})^{-1}$$

we get instead of (2.3.10)

$$(2.3.12) \quad V^{\varkappa,\ell}_{\mu'\mu}(A) = A_{11}^{\mu'+\mu+\varkappa}\,A_{12}^{\mu'-\mu}\times \frac{-1}{2\pi i}\int_{(1^+,0^+)} dt\,(-t)^{\ell+\varkappa+\mu'}(1-t)^{-\ell-\varkappa-1-\mu}(1+A_{21}A_{12}t)^{\ell-\mu'}$$

where the cuts in the t-plane run from $(-A_{12}A_{21})^{-1}$ to $-\infty$ and from 1 to 0. The integration path $(1^+,0^+)$ surrounds the second cut in the positive sense. For the phases we have $\operatorname{arc}(-t) = \operatorname{arc}(1-t) = \operatorname{arc}(1+A_{12}A_{21}t) = 0$ for $(-A_{12}A_{21})^{-1} < t < 0$. Now we can show that the double loop integral

$$(2.3.13) \quad \frac{A_{11}^{\mu+\mu'+\varkappa}A_{12}^{\mu'-\mu}}{4\pi\sin\pi(\mu'+\ell+\varkappa+1)}\int_{(1^+,0^+,1^-,0^-)} dt\; t^{\ell+\varkappa+\mu'}(1-t)^{-\ell-\varkappa-1-\mu}(1+A_{12}A_{21}t)^{\ell-\mu'},$$

the integrand of which is single-valued in the t-plane cut from 0 to $(-A_{12}A_{21})^{-1}$ and from 1 to ∞ with the phases $\operatorname{arc} t = \operatorname{arc}(1-t) = \operatorname{arc}(1+A_{12}A_{21}t) = 0$ for $0<t<1$, coincides with (2.3.12), because the sum of the exponents of t and 1-t in the integrand is an integer. The expression (2.3.13), however, essentially is Pochhammers integral representation of the hypergeometric function $F = {}_2F_1$ (cf. for instance ERDELYI et al. [23]). So we finally get

$$V^{\varkappa,\ell}_{\mu'\mu}(A) = A_{11}^{\mu'+\mu+\varkappa}(-A_{12})^{\mu'-\mu}\,\frac{\Gamma(\mu'+\ell+\varkappa+1)}{\Gamma(\mu+\ell+\varkappa+1)} \times \tag{2.3.14}$$
$$\times\,\frac{F(\mu'-\ell,\mu'+\ell+\varkappa+1;\mu'-\mu+1;-A_{12}A_{21})}{\Gamma(\mu'-\mu+1)} \quad .$$

With (2.3.7) this yields for the matrix elements of unitary representations

$$U^{\varkappa,\ell,\eta}_{SU(1,1)}(A)_{\mu'\mu} = (A_{11}/|A_{11}|)^{\mu'+\mu+\varkappa}(A_{12}/|A_{12}|)^{\mu'-\mu}\,\hat{u}^{\varkappa,\ell}_{\mu'\mu}(-|A_{12}|^2) =$$
$$= (A_{11}/A_{22})^{(\mu'+\mu+\varkappa)/2}(A_{12}/A_{21})^{(\mu'-\mu)/2}\,\hat{u}^{\varkappa,\ell}_{\mu'\mu}(-A_{12}A_{21}), \tag{2.3.15}$$
$$\hat{u}^{\varkappa,\ell}_{\mu'\mu}(x) = \left[\frac{\Gamma(\mu'-\ell)\Gamma(\mu'+\ell+\varkappa+1)}{\Gamma(\mu-\ell)\Gamma(\mu+\ell+\varkappa+1)}\right]^{1/2}(-)^{\mu'-\mu}(-x)^{(\mu'-\mu)/2}(1-x)^{(\mu'+\mu+\varkappa)/2}\times$$
$$\times\,\frac{F(\mu'-\ell,\mu'+\ell+\varkappa+1;\mu'-\mu+1;x)}{\Gamma(\mu'-\mu+1)}\,, \qquad -\infty < x \leq 0 \;.$$

For the derivation of symmetry relations for the matrix elements (2.3.15) we consider the linear operator $T_\varkappa$ and the antilinear operator $K_\varkappa$ defined on $\mathfrak{F}$ by

$$(T_\varkappa f)(\omega) := \omega^\varkappa f(-\omega^{-1})\;, \qquad (K_\varkappa f)(\omega) := \omega^\varkappa f(-\omega)^* \;, \tag{2.3.16}$$
$$T_\varkappa\, e_\nu = (-)^\varkappa\, e_{\varkappa-\nu} = K_\varkappa\, e_\nu \;.$$

They fulfil the conditions

$$T_\varkappa^{\;2} = (-)^\varkappa \mathbb{1}_{\mathfrak{F}} = K_\varkappa^{\;2} \quad . \tag{2.3.17}$$

For the representation $V^{\varkappa,\ell}$ of SU(1,1) on $\mathfrak{F}$ we get

$$T_\varkappa\, V^{\varkappa,\ell}(A)\, T_\varkappa^{-1} = V^{\varkappa,\ell}(A^{-1T}),\quad K_\varkappa\, V^{\varkappa,\ell}(A)\, K_\varkappa^{-1} = V^{\varkappa,\ell^*}(A^{-1\dagger}) \tag{2.3.18}$$

from which with the help of (1.3.32) follows for the matrix elements

$$V^{\varkappa,\ell}_{-\mu'-\varkappa,-\mu-\varkappa}(A^{-1T}) = (-)^{\mu'-\mu}\, V^{\varkappa,\ell}_{\mu'\mu}(A) = V^{\varkappa,-\ell-\varkappa-1}_{-\mu-\varkappa,-\mu'-\varkappa}(A^\dagger) \quad . \tag{2.3.19}$$

Form (2.3.7), (2.3.6) and the equivalence relation (1.3.33) between $V^{\varkappa,\ell}$ and $V^{\varkappa,-\ell-\varkappa-1}$ we get, first for principal and supplementary series only, the symmetry relations

$$(2.3.20)\quad \begin{aligned} &U^{\varkappa,\ell,o}_{SU(1,1)}(A^{-1T})_{\mu'\mu} = U^{\varkappa,\ell,o}_{SU(1,1)}(A)_{-\mu'-\varkappa,-\mu-\varkappa} = U^{\varkappa,\ell,o}_{SU(1,1)}(A^{\dagger})_{\mu\mu'}\,, \\ &(-)^{\mu'-\mu}\,\hat{u}^{\varkappa,\ell}_{\mu'\mu}(x) = \hat{u}^{\varkappa,\ell}_{-\mu'-\varkappa,-\mu-\varkappa}(x) = \hat{u}^{\varkappa,\ell}_{\mu\mu'}(x)\,. \end{aligned}$$

For the derivation of the corresponding relations for the discrete series we have to restrict first the operators $T_\varkappa$ and $K_\varkappa$ to subspaces of $\mathfrak{F}$. The subspaces $\mathfrak{F}^{\varkappa,n}_{+}$ and $\mathfrak{F}^{\varkappa,n}_{-}$ according to (2.3.16) are mapped onto each other by the operators $T_\varkappa$ and $K_\varkappa$. So instead of (2.3.18) we get the relations

$$(2.3.21)\quad \begin{aligned} &\hat{T}_{\pm,\varkappa}\, V^{\varkappa,\ell}_{\pm}(A)\, \hat{T}_{\pm,\varkappa}^{-1} = V^{\varkappa,\ell}_{\mp}(A^{-1T})\,, \\ &\hat{K}_{\pm,\varkappa}\, V^{\varkappa,\ell}_{\pm}(A)\, \hat{K}_{\pm,\varkappa}^{-1} = V^{\varkappa,\ell *}_{\mp}(A^{-1\dagger})\,, \end{aligned}$$

where $\hat{T}_{\pm,\varkappa}$ and $\hat{K}_{\pm,\varkappa}$ are the restrictions to the representation spaces of $V^{\varkappa,\ell}_{\pm}$ of the operators $T_\varkappa$ and $K_\varkappa$. So generalizing (2.3.20) we get for $\eta = 0, \pm$:

$$(2.3.22)\quad \begin{aligned} &U^{\varkappa,\ell,\eta}_{SU(1,1)}(A^{-1T})_{\mu'\mu} = U^{\varkappa,\ell,-\eta}_{SU(1,1)}(A)_{-\mu'-\varkappa,-\mu-\varkappa} = U^{\varkappa,\ell,\eta}_{SU(1,1)}(A^{\dagger})_{\mu\mu'}\,, \\ &(-)^{\mu'-\mu}\,\hat{u}^{\varkappa,\ell}_{\mu'\mu}(x) = \hat{u}^{\varkappa,\ell}_{-\mu'-\varkappa,-\mu-\varkappa}(x) = \hat{u}^{\varkappa,\ell}_{\mu\mu'}(x)\,. \end{aligned}$$

2.4 Matrix Elements of the Irreducible Unitary Representations of E(2) Relative to an H_1-Basis

In this case we may restrict our discussion to the infinite-dimensional representations (1.4.16) of E(2), for the others because of (1.4.13) are identical with the irreducible unitary representations of H_1.

An orthonormal basis of the Hilbert space $\mathfrak{h}^{\rho,\varkappa}_{E(2)}$, according to (1.4.11), is given by

$$(2.4.1)\quad \begin{aligned} &\left\{\psi^{\rho,\varkappa}_{\mu},\ \mu = 0,\pm1,\pm2,\ \dots\ :\ \psi^{\rho,\varkappa}_{\mu}(\zeta) = (\zeta/i\rho)^{-\mu}\right\}, \\ &\langle \psi^{\rho,\varkappa}_{\mu'} | \psi^{\rho,\varkappa}_{\mu}\rangle^{\rho,\varkappa}_{E(2)} = \delta_{\mu'\mu}. \end{aligned}$$

From (1.4.16) follows

$$(2.4.2) \qquad (U^{\rho,\varkappa}_{E(2)}(e^{i\varphi/2},0)\,\psi^{\rho,\varkappa}_{\mu})(\zeta) = \chi^{\varkappa,\mu}(\varphi)\,\psi^{\rho,\varkappa}_{\mu}(\zeta)\ .$$

Therefore $\{\psi^{\rho,\varkappa}_{\mu}\}$ is a basis belonging to the subgroup H_1. An irreducible representation $\chi^{\varkappa',\mu}$ of H_1 appears in $U^{\rho,\varkappa}_{E(2)}$, $\rho > 0$, exactly once for $\varkappa' = \varkappa$, otherwise not.

The matrix elements

$$(2.4.3) \qquad U^{\rho,\varkappa}_{E(2)}(A)_{\mu'\mu} := \langle \psi^{\rho,\varkappa}_{\mu'} | U^{\rho,\varkappa}_{E(2)}(A)\,\psi^{\rho,\varkappa}_{\mu} \rangle^{\rho,\varkappa}_{E(2)}$$

can be expressed easily by an integral which leads to Bessel functions:

$$(2.4.4) \qquad \begin{aligned} U^{\rho,\varkappa}_{E(2)}(A)_{\mu'\mu} &= (A_{11}/|A_{11}|)^{\mu'+\mu+\varkappa}(A_{12}/|A_{12}|)^{\mu'-\mu}\, u^{\rho,\varkappa}_{\mu'\mu}(|A_{12}|^2)\ , \\ u^{\rho,\varkappa}_{\mu'\mu}(z) &:= J_{\mu'-\mu}(\rho\sqrt{z})\ , \quad 0 \le z < \infty\ . \end{aligned}$$

The symmetry relations

$$(2.4.5) \qquad (-)^{\mu'-\mu}\, u^{\rho,\varkappa}_{-\mu'-\varkappa,-\mu-\varkappa}(z) = u^{\rho,\varkappa}_{\mu'\mu}(z) = u^{\rho,\varkappa}_{-\mu-\varkappa,-\mu'-\varkappa}(z)$$

at once follow from the Bessel function relation $J_{-n} = (-)^n J_n$. For the matrix elements $U^{\rho,\varkappa}_{E(2)}(A)_{\mu'\mu}$ we therefore get

$$(2.4.6) \qquad (-)^{\mu'-\mu} U^{\rho,\varkappa}_{E(2)}(A^*)_{\mu'\mu} = U^{\rho,\varkappa}_{E(2)}(A)_{-\mu'-\varkappa,-\mu-\varkappa} = (-)^{\mu'-\mu} U^{\rho,\varkappa}_{E(2)}(A^{-1})_{\mu\mu'}\ .$$

2.5 Matrix Elements of the Irreducible Unitary Representations of SU(1,1) Relative to an H_2-Basis

We begin with a purely formal projection technic as a guide to the solution of the reduction problem for the restriction of the irreducible unitary representations of SU(1,1) to the noncompact subgroup $H_2 = SU(1,1) \cap SL(2,\mathbb{R})$. The operator

$$(2.5.1) \qquad P^{\varkappa,\ell,\eta}_{\lambda} := \frac{1}{2}\sum_{\varepsilon=\pm} \int_{-\infty}^{+\infty} \frac{d\xi}{2\pi}\, \chi^{\varkappa,\lambda}(\varepsilon,\xi)\, U^{\varkappa,\ell,\eta}_{SU(1,1)}(D(\varepsilon,\xi))$$

has the projection properties

$$(2.5.2) \qquad P^{\varkappa,\ell,\eta\,\dagger}_{\lambda} = P^{\varkappa,\ell,\eta}_{\lambda}\ , \quad P^{\varkappa,\ell,\eta}_{\lambda'}\, P^{\varkappa,\ell,\eta}_{\lambda} = \delta(\lambda'-\lambda)\, P^{\varkappa,\ell,\eta}_{\lambda}\ .$$

Because of

$$(2.5.3) \qquad U^{\varkappa,\ell,\eta}_{SU(1,1)}(D(\varepsilon,\xi))\, P^{\varkappa,\ell,\eta}_{\lambda} = \chi^{\varkappa,\lambda}(\varepsilon,\xi)\, P^{\varkappa,\ell,\eta}_{\lambda}$$

we can interpret $P^{\varkappa,\ell,\eta}_{\lambda}$ as projecting out a multiple of the irreducible unitary representation $\chi^{\varkappa,\lambda}$ of H_2 from $U^{\varkappa,\ell,\eta}_{SU(1,1)}|H_2$. From (2.5.1) we have

$$(P^{\varkappa,\ell,\eta}_{\lambda} f)(\omega) = \int_{-\infty}^{+\infty} \frac{d\xi}{2\pi} e^{-i\lambda\xi} |\omega \sinh\xi/2 + \cosh\xi/2|^{-2\ell-2\varkappa-2} \times$$

$$\times (\omega \sinh\xi/2 + \cosh\xi/2)^{\varkappa} f(\omega\overline{D(\varepsilon,\xi)}) =$$

$$= \sum_{\tau=\pm} \varphi^{\varkappa,\ell}_{\tau\lambda}(\;)\, f''_{\tau}(\lambda) \quad ,$$

$$(2.5.4) \qquad f''_{\tau}(\lambda) := \frac{1}{2\pi i} \oint \frac{d\omega}{\omega} \varphi^{\varkappa,-\ell^*-\varkappa-1}_{\tau\lambda}(\omega)\, f(\omega) \quad ,$$

$$\varphi^{\varkappa,\ell}_{\tau\lambda}(\omega) := \begin{cases} \omega^{\varkappa/2} \left|\frac{\omega-\omega^{-1}}{2}\right|^{-\ell-\varkappa/2-1} \left|\frac{1-\omega}{1+\omega}\right|^{-i\lambda} & \text{for sign Im}\omega = \tau \,, \\ 0 & \text{otherwise} \,, \end{cases}$$

$$\omega^{\varkappa/2} := e^{i\varkappa(\mathrm{arc}\,\omega)/2} \,, \quad -\pi \le \mathrm{arc}\,\omega < \pi \,.$$

For the substitution

$$(2.5.5) \qquad \xi \longrightarrow \omega' = \omega\overline{D(\varepsilon,\xi)}$$

used therein one should note that the boundary ∂K of the unit circle has the following partition into orbits with respect to H_2:

$$(2.5.6) \qquad \begin{aligned} \partial K &:= \{\omega \in \mathbb{C} : |\omega| = 1\} = \partial K_+ \cup \partial K_- \cup \{1\} \cup \{-1\} \,, \\ \partial K_\tau &:= \{\omega \in \partial K : \text{sign Im}\,\omega = \tau\} \,, \quad \tau = \pm \,. \end{aligned}$$

From this partition we see that any subspace of $\mathfrak{H}^{\varkappa,\ell,\eta}_{SU(1,1)}$, consisting of functions which are concentrated to one of the orbits, is invariant under $U^{\varkappa,\ell,\eta}_{SU(1,1)}|H_2$. The orbits $\{1\}$ and $\{-1\}$, however, as subsets of measure zero in ∂K may be neglected in the following.

The exact solution of the reduction problem will be given at first for the principal series. In a first step we unitarily map the representation space $\mathfrak{h}^{\varkappa,\ell,0}_{SU(1,1)}$ by

$$(2.5.7)\quad \begin{aligned} f'_\tau(q) &:= (\cosh q)^{-\ell-\varkappa/2-1}\left(\frac{1+i\tau e^{-q}}{1-i\tau e^{-q}}\right)^{-\varkappa/2} f\left(\frac{1+i\tau e^{-q}}{1-i\tau e^{-q}}\right), \\ &\frac{1+i\tau e^{-q}}{1-i\tau e^{-q}} \in \partial K , \quad q \in \mathbb{R} , \end{aligned}$$

onto the Hilbert space $\mathfrak{h}'^{\varkappa,\ell,0}_{SU(1,1)}$ consisting of pairs $f' = (f'_+, f'_-)$ of functions defined on $\mathbb{R}$ with scalar product

$$(2.5.8)\quad \langle f'|g'\rangle'^{\varkappa,\ell,0}_{SU(1,1)} := \sum_{\tau=\pm}\int_{-\infty}^{+\infty}\frac{dq}{2\pi}\, f'_\tau(q)^* g'_\tau(q) = \langle f|g\rangle^{\varkappa,\ell,0}_{SU(1,1)} .$$

The Fourier transformation

$$(2.5.9)\quad f''_\tau(\lambda) = \int_{-\infty}^{+\infty}\frac{dq}{2\pi}\, e^{-i\lambda q} f'_\tau(q) , \quad f'_\tau(q) = \int_{-\infty}^{+\infty} d\lambda\, e^{i\lambda q} f''_\tau(\lambda)$$

yields another unitary map of $\mathfrak{h}'^{\varkappa,\ell,0}_{SU(1,1)}$ onto the Hilbert space $\mathfrak{h}''^{\varkappa,\ell,0}_{SU(1,1)}$ consisting of pairs $f'' = (f''_+, f''_-)$ of functions defined on $\mathbb{R}$ with scalar product

$$(2.5.10)\quad \langle f''|g''\rangle''^{\varkappa,\ell,0}_{SU(1,1)} := \sum_{\tau=\pm}\int_{-\infty}^{+\infty} d\lambda\, f''_\tau(\lambda)^* g''_\tau(\lambda) = \langle f|g\rangle^{\varkappa,\ell,0}_{SU(1,1)} .$$

We may equip $\mathfrak{h}''^{\varkappa,\ell,0}_{SU(1,1)}$ with the structure

$$(2.5.11)\quad \mathfrak{h}''^{\varkappa,\ell,0}_{SU(1,1)} = \oplus\int_{-\infty}^{+\infty}\sqrt{d\lambda}\; \mathbb{C}^2$$

of a direct integral over $\mathbb{R}$ of two-dimensional complex Hilbert spaces. Carrying over the representation $U^{\varkappa,\ell,0}_{SU(1,1)}|H_2$ to $\mathfrak{h}''^{\varkappa,\ell,0}_{SU(1,1)}$ we get

$$(2.5.12)\quad (U''^{\varkappa,\ell,0}_{SU(1,1)}(D(\varepsilon,\xi))f''_\tau)(\lambda) = \chi^{\varkappa,\lambda}(\varepsilon,\xi)\, f''_\tau(\lambda) .$$

This solves completely the reduction problem for $U^{\varkappa,\ell,0}_{SU(1,1)}|H_2$ because $U''^{\varkappa,\ell,0}_{SU(1,1)}|H_2$ has the structure

$$(2.5.13)\quad U''^{\varkappa,\ell,0}_{SU(1,1)}|H_2 = \oplus\int_{-\infty}^{+\infty} d\lambda\, (\chi^{\varkappa,\lambda} \oplus \chi^{\varkappa,\lambda})$$

of a direct integral of multiples of irreducible unitary representations $\chi^{\varkappa',\lambda}$ of H_2, each occurring exactly twice if $\varkappa' = \varkappa$, otherwise not. This is in accordance with BARGMANNs [13] assertion concerning the multiplicity of the spectrum of the generator belonging to the subgroup H_2. The transition from $\mathfrak{h}^{\varkappa,\ell,o}_{SU(1,1)}$ to $\mathfrak{h}''^{\varkappa,\ell,o}_{SU(1,1)}$ which solves the reduction problem, was inspired by the heuristic consideration at the beginning of this Section, for obviously the Fourier transform f''_τ from (2.5.9) coincides with the function f''_τ in (2.5.4) which according to (2.5.3) formally satisfies the representation formula (2.5.12). The functions $\varphi^{\varkappa,\ell}_{\tau\lambda}$ defined in (2.5.4) play the role of integral kernels for the unitary map between $\mathfrak{h}^{\varkappa,\ell,o}_{SU(1,1)}$ and $\mathfrak{h}''^{\varkappa,\ell,o}_{SU(1,1)}$. The unitarity of this map may be expressed by the relations

$$\langle \varphi^{\varkappa,\ell}_{\tau'\lambda'} | \varphi^{\varkappa,\ell}_{\tau\lambda} \rangle^{\varkappa,\ell,o}_{SU(1,1)} = \delta_{\tau'\tau}\,\delta(\lambda'-\lambda) \,,$$

(2.5.14)

$$\sum_{\tau',\tau=\pm} \int_{-\infty}^{+\infty} d\lambda \; \varphi^{\varkappa,\ell}_{\tau\lambda}(\omega')\, \varphi^{\varkappa,\ell}_{\tau\lambda}(\omega)^* = 2\pi\,\delta(-i\ln\omega'/\omega) \quad .$$

For the supplementary series we may try the transformation

$$f''_\tau(\lambda) = \frac{1}{2\pi i}\oint \frac{d\omega}{\omega}\,\varphi^{o,-\ell-1}_{\tau\lambda}(\omega)^* f(\omega) \quad ,$$

(2.5.15)

$$f(\omega) = \sum_{\tau=\pm} \int_{-\infty}^{+\infty} d\lambda\; \varphi^{o,\ell}_{\tau\lambda}(\omega)\, f''_\tau(\lambda) \quad ,$$

which is inspired by (2.5.4) and summarizes the transformations analogous to (2.5.7) and (2.5.9). This is a unitary map between $\mathfrak{h}^{o,\ell,o}_{SU(1,1)}$ and the Hilbert space $\mathfrak{h}''^{o,\ell,o}_{SU(1,1)}$ of pairs $f'' = (f''_+, f''_-)$ of functions on the real line with scalar product

$$\langle f''|g''\rangle''^{o,\ell,o}_{SU(1,1)} := \int_{-\infty}^{+\infty} d\lambda \sum_{\tau',\tau=\pm} f''_{\tau'}(\lambda)^*\, T''^{o,\ell}_{\tau'\tau}(\lambda)\, g''_\tau(\lambda) = \langle f|g\rangle^{o,\ell,o}_{SU(1,1)} ,$$

(2.5.16)

$$T''^{o,\ell}_{\tau'\tau}(\lambda) := \frac{|\Gamma(-\ell+i\lambda)|^2}{\pi}\,(\delta_{\tau'\tau}\cosh\pi\lambda + \delta_{\tau',-\tau}\cos\pi\ell) \;.$$

The form of the scalar product in $\mathfrak{h}''^{o,\ell,o}_{SU(1,1)}$ follows from the integral formula

$$\frac{1}{2\pi i}\oint \frac{d\omega'}{\omega'}\, T^{\varkappa,\ell}(\omega/\omega')\,\varphi^{\varkappa,\ell}_{\tau\lambda}(\omega') = \sum_{\tau'=\pm} \varphi^{\varkappa,-\ell-\varkappa-1}_{\tau'\lambda}(\omega)\; T''^{\varkappa,\ell}_{\tau'\tau}(\lambda) \;,$$

(2.5.17)

$$T''^{\varkappa,\ell}(\lambda) := \frac{\Gamma(-\ell-\varkappa/2-i\lambda)\Gamma(-\ell-\varkappa/2+i\lambda)}{\pi}\, i^{\varkappa} \begin{pmatrix} \cos\pi(i\lambda-\varkappa/2) & (-)^{\varkappa}\cos\pi\ell \\ \cos\pi\ell & \cos\pi(i\lambda+\varkappa/2) \end{pmatrix}$$

which holds for ℓ from the strip $\{z \in \mathbb{C}:\ -1 < \mathrm{Re}(z+\varkappa/2) < 0\}$. Obviously $\mathfrak{h}''^{o,\ell,o}_{SU(1,1)}$ may be given the direct integral stnucture

(2.5.18)
$$\mathfrak{h}''^{o,\ell,o}_{SU(1,1)} = \oplus\int_{-\infty}^{+\infty} \sqrt{d\lambda}\; \mathbb{C}^2_\ell(\lambda) \;,$$

where $\mathbb{C}^2_\ell(\lambda)$ is the two-dimensional complex Hilbert space with the metric defined by $T''^{0,\ell}(\lambda)$. The restricted representation $U^{0,\ell,0}_{SU(1,1)}|H_2$, when carried over to $\mathfrak{h}''^{0,\ell,0}_{SU(1,1)}$, again has the form given by (2.5.12) and (2.5.13) so that the reduction problem for the supplementary series finds the same solution as for the principal series. The unitarity relations corresponding to (2.5.14) are

$$\langle \varphi^{0,\ell}_{\tau'\lambda'} | \varphi^{0,\ell}_{\tau\lambda} \rangle^{0,\ell,0}_{SU(1,1)} = T''^{0,\ell}_{\tau'\tau}(\lambda)\, \delta(\lambda'-\lambda) \quad ,$$

(2.5.19)

$$\sum_{\tau',\tau=\pm} \int_{-\infty}^{+\infty} d\lambda\, \varphi^{0,\ell}_{\tau'\lambda}(\omega')\, T''^{0,-\ell-1}_{\tau'\tau}(\lambda)\, \varphi^{0,\ell}_{\tau\lambda}(\omega)^* = T^{0,-\ell-1}(\omega'/\omega) \ .$$

For the discrete series at first we modify the heuristic consideration which leads to (2.5.4). From the construction of the realization (1.3.46) of this series follows that the elements of the representation spaces $\mathfrak{h}^{\varkappa,\ell,\eta}_{SU(1,1)}$, $\eta = \pm$, may be considered as boundary values of functions which are analytic in the interior or in the exterior of the unit circle. If f in (2.5.4) has such a boundary property it is unconvenient to use the integral kernels $\varphi^{\varkappa,\ell}_{\tau\lambda}(\omega)$ which are concentrated only to one half of the boundary of the unit circle. We therefore construct linear combinations of the $\varphi^{\varkappa,\ell}_{\tau\lambda}(\omega)$ which a priori reflect the behaviour of the f. By

$$w^{\varkappa,\ell}_{+,\lambda}(z) := \frac{\Gamma(1+\ell+\varkappa/2+i\lambda)}{\sqrt{2\pi}} \left(\frac{1-z}{\sqrt{2}}\right)^{-1-\ell-\varkappa/2-i\lambda} \left(\frac{1+z}{\sqrt{2}}\right)^{-1-\ell-\varkappa/2+i\lambda} \quad ,$$

(2.5.20)

$$w^{\varkappa,\ell}_{-,\lambda}(z) := z^{\varkappa}\, w^{\varkappa,\ell}_{+,\lambda}(1/z) \quad ,$$

$$z^{-\ell-\varkappa-1}\, w^{\varkappa,\ell}_{+,\lambda}(z)\Big|_{z=0} = 2^{1+\ell+\varkappa/2}\, \frac{\Gamma(1+\ell+\varkappa/2+i\lambda)}{\sqrt{2\pi}} \quad ,$$

we define two functions which are single-valued and analytic in the interior and the exterior, respectively, of the unit circle and on $\partial K_+ \cup \partial K_-$ assume the values

$$w^{\varkappa,\ell}_{\tau\lambda}(\omega) := \lim_{z\to\omega} w^{\varkappa,\ell}_{\tau\lambda}(z) =$$

(2.5.21)

$$= \frac{\Gamma(1+\ell+\varkappa/2+i\lambda)}{\sqrt{2\pi}} \sum_{\tau'=\pm} e^{i\tau'\tau(1+\ell+\varkappa/2+i\lambda)\pi/2}\, \varphi^{\varkappa,\ell}_{\tau'\lambda}(\omega) \ .$$

Clearly $w^{\varkappa,\ell}_{+,\lambda}$ is a single-valued analytical function in the complex z-plane cut from $-\infty$ to -1 and from 1 to $+\infty$ along the real axis. If we use the linear combinations $w^{\varkappa,\ell}_{\tau\lambda}(\omega)$ instead of the $\varphi^{\varkappa,\ell}_{\tau\lambda}(\omega)$ in (2.5.4) we get

$$(2.5.22)\qquad \begin{aligned}(P_\lambda^{\varkappa,\ell,\eta}f)(\omega) &= \sum_{\tau=\pm} w_{\tau\lambda}^{\varkappa,\ell}(\omega)\,\tilde f_\tau(\lambda)\ ,\\ \tilde f_\tau(\lambda) &= \frac{1}{2\pi i}\oint\frac{d\omega}{\omega}\,w_{\tau\lambda}^{\varkappa,-\ell-\varkappa-1}(\omega)^*\,f(\omega)\ .\end{aligned}$$

The condition of f being the boundary value of an analytical function in the interior ($\eta = +$) or the exterior ($\eta = -$) of the unit circle may be expressed by the two equivalent conditions

$$(2.5.23)\qquad (P_\lambda^{\varkappa,\ell,\eta}f)(\omega) = w_{\eta\lambda}^{\varkappa,\ell}(\omega)\,\tilde f_\eta(\lambda)\ ,\qquad \tilde f_{-\eta}(\lambda) \equiv 0\ .$$

We still may modify somewhat the integral in (2.5.22) which defines $\tilde f_\eta(\lambda)$. At first we continue analytically the relation (2.5.17) into the whole complex ℓ-plane with exception of the points $\pm i\lambda-\varkappa/2+n$, $n = 0,1,2,\ \ldots$, where the Γ-function in $T''^{\varkappa,\ell}(\lambda)$ has poles, and the points $\pm i\lambda-\varkappa/2-1-n$, $n = 0,1,2,\ \ldots$, where the matrix $T''^{\varkappa,\ell}(\lambda)$ degenerates. Up to these exceptional points for all complex ℓ holds the relation

$$(2.5.24)\qquad \begin{aligned}\sum_{\tau''=\pm} T''^{\varkappa,\ell}_{\tau'\tau''}(\lambda)e^{i\tau''\tau(1+\ell+\varkappa/2+i\lambda)\pi/2} &= \frac{\tau^\varkappa\Gamma(-\ell-\varkappa/2+i\lambda)}{\Gamma(1+\ell+\varkappa/2+i\lambda)}\ \times\\ \times\, e^{i\tau'\tau(-\ell-\varkappa/2+i\lambda)\pi/2}\ , &\quad \ell \neq \pm i\lambda-\varkappa/2+n,\ n = 0,\pm1,\pm2,\ \ldots\ .\end{aligned}$$

With (2.5.21) for integer $\ell \geq 0$ we therefore get for the operator $\hat T_\eta^{\varkappa,\ell}$, which according to (1.3.30) is restricted to $\mathfrak{F}_\eta^{\varkappa,\ell}$, the expression

$$(2.5.25)\qquad (\eta^\varkappa\,\hat T_\eta^{\varkappa,\ell}\ w_{\eta\lambda}^{\varkappa,\ell})(\omega) = (P_\eta^{\varkappa,\ell}\ w_{\eta\lambda}^{\varkappa,-\ell-\varkappa-1})(\omega)\ .$$

Here $P_\eta^{\varkappa,\ell}$ means the projector onto $\mathfrak{F}_\eta^{\varkappa,\ell}$. With this we may express $\tilde f_\eta(\lambda)$ by

$$(2.5.26)\qquad \tilde f_\eta(\lambda) = \left\langle w_{\eta\lambda}^{\varkappa,\ell}\middle| f\right\rangle_{SU(1,1)}^{\varkappa,\ell,\eta}$$

with the scalar product in $\mathfrak{h}_{SU(1,1)}^{\varkappa,\ell,\eta}$ defined in (1.3.41). The exact solution of the reduction problem for the discrete series will be presented only for the case $\eta = +$. The other part of this series according to (1.3.49) may be realized by $U_{SU(1,1)}^{|\varkappa,\ell,+\,*}$. We use the results of Appendix A. It is shown there that the functions

$$(2.5.27)\quad K_{\mu}^{\varkappa,\ell}(\lambda) := \frac{\Gamma(1+\ell+\varkappa/2-i\lambda)}{\sqrt{2\pi}}\, 2^{1+\ell+\varkappa/2}\left[\frac{(\mu+\ell+\varkappa)!}{(\mu-\ell-1)!}\right]^{1/2}\frac{1}{\Gamma(2\ell+\varkappa+2)}\times$$
$$\times F(1+\ell-\mu, 1+\ell+\varkappa/2+i\lambda; 2\ell+\varkappa+2; 2),\quad \mu = \ell+1, \ell+2, \ldots ,$$

form a complete orthonormal basis in the Hilbert space $\tilde{\mathfrak{h}}_+ = \mathcal{L}^2(\mathbb{R})$. According to (A.5) $w_{+,\lambda}^{\varkappa,\ell}(z)$ has the expansion

$$(2.5.28)\qquad w_{+,\lambda}^{\varkappa,\ell}(z) = \sum_{\mu=\ell+1} \psi_{\mu}^{\prime\varkappa,\ell,+}(z)\, K_{\mu}^{\varkappa,\ell}(\lambda)^*$$

with the orthonormal basis $\{\psi_{\mu}^{\prime\varkappa,\ell,+}\}$ of the representation space $\mathfrak{h}_{SU(1,1)}^{\prime\varkappa,\ell,+}$. Therefore $w_{+,\lambda}^{\varkappa,\ell}(z)$ forms the integral kernel of a unitary map between $\mathfrak{h}_{SU(1,1)}^{\prime\varkappa,\ell,+}$ and $\tilde{\mathfrak{h}}_+$ with the transformation formulas

$$(2.5.29)\quad \tilde{f}_+(\lambda) = \int_{|z|<1} d\mu^{\varkappa,\ell}(z)\, w_{+,\lambda}^{\varkappa,\ell}(z)^* f'(z),\quad f'(z) = \int_{-\infty}^{+\infty} d\lambda\, w_{+,\lambda}^{\varkappa,\ell}(z)\, \tilde{f}_+(\lambda),$$

which correspond to (2.5.26) and (2.5.23). If we carry over the representations of the discrete series to $\tilde{\mathfrak{h}}_+$ we get

$$(2.5.30)\qquad (\tilde{U}_{SU(1,1)}^{\varkappa,\ell,+}(D(\varepsilon,\xi))\tilde{f}_+)(\lambda) = \chi^{\varkappa,\lambda}(\varepsilon,\xi)\, \tilde{f}_+(\lambda)$$

so that for $\tilde{U}_{SU(1,1)}^{\varkappa,\ell,+}\big|H_2$ we have the decomposition

$$(2.5.31)\qquad \tilde{\mathfrak{h}}_+ = \oplus\int_{-\infty}^{+\infty} \sqrt{d\lambda}\, \mathbb{C}\ ,\quad \tilde{U}_{SU(1,1)}^{\varkappa,\ell,+}\big|H_2 = \oplus\int_{-\infty}^{+\infty} d\lambda\, \chi^{\varkappa,\lambda}\ .$$

Therefore an irreducible unitary representation $\chi^{\varkappa',\lambda}$ of H_2 is contained in $U_{SU(1,1)}^{\prime\varkappa,\ell,+}\big|H_2$ exactly once if $\varkappa' = \varkappa$, and otherwise not. The unitarity of the transformation performed with the $w_{+,\lambda}^{\varkappa,\ell}(z)$ is expressed by the relations (cf. also (A.8))

$$(2.5.32)\qquad \begin{aligned} \int_{|z|<1} d\mu^{\varkappa,\ell}(z)\, w_{+,\lambda'}^{\varkappa,\ell}(z)^* w_{+,\lambda}^{\varkappa,\ell}(z) &= \delta(\lambda'-\lambda)\quad , \\ \int_{-\infty}^{+\infty} d\lambda\, w_{+,\lambda}^{\varkappa,\ell}(z')\, w_{+,\lambda}^{\varkappa,\ell}(z)^* &= K^{\varkappa,\ell,+}(z',z), \end{aligned}$$

where $K^{\varkappa,\ell,+}$ is the reproducing kernel (1.3.52) in $\mathfrak{h}_{SU(1,1)}^{\prime\varkappa,\ell,+}$. To these correspond the unitarity relations

$$(2.5.33)\qquad \begin{aligned} \frac{1}{2\pi i}\oint\frac{d\omega}{\omega}\, w_{\eta\lambda'}^{\varkappa,\ell}(\omega)^*\,(\eta^{\varkappa}\,\hat{T}_{\eta}^{\varkappa,\ell}\, w_{\eta\lambda}^{\varkappa,\ell})(\omega) &= \delta(\lambda'-\lambda)\quad , \\ \int_{-\infty}^{+\infty} d\lambda\, w_{\eta\lambda}^{\varkappa,\ell}(\omega')\, w_{\eta\lambda}^{\varkappa,\ell}(\omega)^* &= \eta^{\varkappa}\hat{T}_{\eta}^{\varkappa,-\ell-\varkappa-1}(\omega'/\omega) \end{aligned}$$

for the transformation performed with the boundary values (2.5.21) between the representation spaces $\mathfrak{h}^{\varkappa,\ell,\eta}_{SU(1,1)}$ and $\tilde{\mathfrak{h}}_\eta = \mathcal{L}^2(\mathbb{R})$.

A unified formulation of the solution of the reduction problem for $U^{\varkappa,\ell,\eta}_{SU(1,1)}\big| H_2$ can be attained by an appropriate unitary map of the Hilbert spaces $\mathfrak{h}''^{\varkappa,\ell,o}_{SU(1,1)}$ and $\tilde{\mathfrak{h}}_+ \oplus \tilde{\mathfrak{h}}_-$, respectively, onto the Hilbert space

(2.5.34) $$\tilde{\mathfrak{h}} = \oplus\int_{-\infty}^{+\infty}\sqrt{d\lambda}\; \mathbb{C}^2, \langle \tilde{f}|\tilde{g}\rangle^{\sim} := \sum_{\tau=\pm}\int_{-\infty}^{+\infty} d\lambda\; \tilde{f}_\tau(\lambda)^* \tilde{g}_\tau(\lambda),\; \tilde{f}, \tilde{g} \in \tilde{\mathfrak{h}} .$$

The 2x2-matrix defined by

(2.5.35) $$N^{\varkappa,\ell}(\lambda) := \frac{\Gamma(1+\ell+\varkappa/2+i\lambda)}{\sqrt{2\pi}}\begin{pmatrix} e^{i(1+\ell+\frac{\varkappa}{2}+i\lambda)\frac{\pi}{2}} & e^{-i(1+\ell+\frac{\varkappa}{2}+i\lambda)\frac{\pi}{2}} \\ e^{-i(1+\ell+\frac{\varkappa}{2}+i\lambda)\frac{\pi}{2}} & e^{i(1+\ell+\frac{\varkappa}{2}+i\lambda)\frac{\pi}{2}} \end{pmatrix}$$

because of

(2.5.36) $$\det N^{\varkappa,\ell}(\lambda) = i\,\Gamma(1+\ell+\varkappa/2+i\lambda)/\Gamma(-\ell-\varkappa/2-i\lambda)$$

is nonsingular for every complex ℓ with the exception of the points $-i\lambda-\varkappa/2+n$, $n = 0,\pm1,\pm2, \ldots$. According to (2.5.24) we therefore have

(2.5.37) $$T''^{\varkappa,\ell}(\lambda)\; N^{\varkappa,\ell}(\lambda) = N^{\varkappa,-\ell-\varkappa-1}(\lambda)(\sigma_3)^{\varkappa},\quad \ell \neq \pm i\lambda-\varkappa/2+n, n=0,\pm1,\ldots$$

For the ℓ-values of the principal series $N^{\varkappa,\ell}(\lambda)$ is unitary:

(2.5.38) $$N^{\varkappa,\ell}(\lambda)^{-1\dagger} = N^{\varkappa,\ell}(\lambda)\quad ,\quad \ell \in -(1+\varkappa)/2 + i\,\mathbb{R} ,$$

while for the ℓ-values of the supplementary series with (2.5.37) we get

(2.5.39) $$N^{o,\ell}(\lambda)^{-1\dagger} = N^{o,-\ell-1}(\lambda) = T''^{o,\ell}(\lambda)\; N^{o,\ell}(\lambda)\quad ,\quad \ell \in (-1,0) .$$

Therefore

(2.5.40) $$f''(\lambda) = N^{\varkappa,\ell}(\lambda)\; \tilde{f}(\lambda)$$

defines a unitary map of the representation spaces $\mathfrak{h}''^{\varkappa,\ell,o}_{SU(1,1)}$ of principal and supplementary series with the scalar products (2.5.10) and (2.5.16) onto the Hilbert space $\tilde{\mathfrak{h}}$ in (2.5.34). For simplification here and in the following we use a matrix notation: f'' and $\tilde{f}$ are to be understood as two-component column vectors. The composed unitary transformation between the representation spaces $\mathfrak{h}^{\varkappa,\ell,o}_{SU(1,1)}$ and $\tilde{\mathfrak{h}}$ may be summarized in the formulas

$$(2.5.41)\quad \tilde{f}_\tau(\lambda) = \langle w^{\varkappa,\ell}_{\tau\lambda} | f \rangle^{\varkappa,\ell,o}_{SU(1,1)}, \quad f(\omega) = \sum_{\tau=\pm} \int_{-\infty}^{+\infty} d\lambda \; w^{\varkappa,\ell}_{\tau\lambda}(\omega)\, \tilde{f}_\tau(\lambda)$$

with the new integral kernels

$$(2.5.42)\quad w^{\varkappa,\ell}_{\tau\lambda}(\omega) := \sum_{\tau'=\pm} \varphi^{\varkappa,\ell}_{\tau'\lambda}(\omega)\, N^{\varkappa,\ell}_{\tau'\tau}(\lambda) .$$

Instead of (2.5.14) and (2.5.19) for them hold the unitarity relations

$$\langle w^{\varkappa,\ell}_{\tau'\lambda'} | w^{\varkappa,\ell}_{\tau\lambda} \rangle^{\varkappa,\ell,o}_{SU(1,1)} = \delta_{\tau'\tau}\delta(\lambda'-\lambda) ,$$

$$(2.5.43)\quad \sum_{\tau=\pm} \int_{-\infty}^{+\infty} d\lambda \; w^{\varkappa,\ell}_{\tau\lambda}(\omega')\, w^{\varkappa,\ell}_{\tau\lambda}(\omega)^* = \begin{cases} 2\pi\, \delta(-i \ln\omega'/\omega), & \ell \in -(1+\varkappa/2)+i\mathbb{R}, \\ T^{o,-\ell-1}(\omega'/\omega), & \ell \in (-1,0) \end{cases} .$$

The integral formula (2.5.17) because of (2.5.37) is changed to

$$(2.5.44)\quad (T^{\varkappa,\ell}\, w^{\varkappa,\ell}_{\tau\lambda})(\omega) = \tau^{\varkappa}\, w^{\varkappa,-\ell-\varkappa-1}_{\tau\lambda}(\omega) .$$

For the integer ℓ-values of the discrete series the expressions (2.5.42) obviously coincide with the boundary values (2.5.21) of the analytical functions defined in (2.5.20). As shown above these constitute unitary maps between the representation spaces $\mathfrak{h}'^{\varkappa,\ell,\eta}_{SU(1,1)}$ and the Hilbert spaces $\tilde{\mathfrak{h}}_\eta \equiv \mathcal{L}^2(\mathbb{R})$. These maps can be combined to a unitary map between $\mathfrak{h}'^{\varkappa,\ell,+}_{SU(1,1)} \oplus \mathfrak{h}'^{\varkappa,\ell,-}_{SU(1,1)}$ and

$$(2.5.45)\quad \tilde{\mathfrak{h}} = \tilde{\mathfrak{h}}_+ \oplus \tilde{\mathfrak{h}}_- = \{(\tilde{f}_+,\tilde{f}_-) : \tilde{f}_\eta \in \tilde{\mathfrak{h}}_\eta\} .$$

For the discrete series $\tilde{\mathfrak{h}}$ therefore appears as representation space of the unitarily transformed representation $U^{\varkappa,\ell,+}_{SU(1,1)} \oplus U^{\varkappa,\ell,-}_{SU(1,1)}$. Hence for the discrete series the formula (2.5.41) may be completed by

$$(2.5.46)\quad \begin{aligned} &\tilde{f}_\eta(\lambda) = \langle w^{\varkappa,\ell}_{\eta\lambda} | f \rangle^{\varkappa,\ell,\eta}_{SU(1,1)} , \quad \tilde{f}_{-\eta}(\lambda) \equiv 0 , \quad f \in \mathfrak{h}^{\varkappa,\ell,\eta}_{SU(1,1)} , \\ &f(\omega) = \sum_{\tau=\pm} \int_{-\infty}^{+\infty} d\lambda \; w^{\varkappa,\ell}_{\tau\lambda}(\omega)\, \tilde{f}_\tau(\lambda) = \int_{-\infty}^{+\infty} d\lambda \; w^{\varkappa,\ell}_{\eta\lambda}(\omega)\, \tilde{f}_\eta(\lambda) , \end{aligned}$$

with the unitarity relations (2.5.33).

The representation $U^{\varkappa,\ell,\eta}_{SU(1,1)}$, when carried over to the Hilbert space $\tilde{\mathfrak{h}}$, according to (2.5.41) and (2.5.46) may be written in the form

$$(2.5.47)\quad (U^{\varkappa,\ell,\eta}_{SU(1,1)}(A)f)^{\sim}_{\tau'}(\lambda') = \sum_{\tau=\pm}\int_{-\infty}^{+\infty} d\lambda\, U^{\varkappa,\ell,\eta}_{SU(1,1)}(A)_{\tau'\lambda',\tau\lambda}\,\tilde{f}_{\tau}(\lambda)$$

with the generalized integral kernels

$$(2.5.48)\quad U^{\varkappa,\ell,\eta}_{SU(1,1)}(A)_{\tau'\lambda',\tau\lambda} := \langle w^{\varkappa,\ell}_{\tau'\lambda'} | U^{\varkappa,\ell,\eta}_{SU(1,1)}(A) w^{\varkappa,\ell}_{\tau\lambda}\rangle^{\varkappa,\ell,\eta}_{SU(1,1)} .$$

Because of

$$(2.5.49)\quad U^{\varkappa,\ell,\eta}_{SU(1,1)}(D(\varepsilon,\xi))_{\tau'\lambda',\tau\lambda} = \chi^{\varkappa,\lambda}(\varepsilon,\xi)\,\delta_{\tau'\tau}\,\delta(\lambda'-\lambda)$$

we denote them as matrix elements of $U^{\varkappa,\ell,\eta}_{SU(1,1)}$ relative to a basis belonging to the subgroup H_2. Obviously the unitarity relations (2.5.43) and (2.5.33) play the role of generalized orthogonality and completeness relations for the "basis" $\{w^{\varkappa,\ell}_{\tau\lambda} : \tau = \pm, -\infty < \lambda < \infty\}$.

We first calculate the generalized matrix elements (2.5.48) only for principal and supplementary series. With (2.5.42), (2.5.44), (2.5.38) and (2.5.39) we get

$$U^{\varkappa,\ell,\eta}_{SU(1,1)}(A)_{\lambda'\lambda} = N^{\varkappa,\ell}(\lambda')^{-1}\, V^{\varkappa,\ell}_{\lambda'\lambda}(A)\, N^{\varkappa,\ell}(\lambda) ,$$

$$(2.5.50)\quad V^{\varkappa,\ell}_{\tau'\lambda',\tau\lambda}(A) = \frac{1}{2\pi i}\oint \frac{d\omega}{\omega}\,\omega^{-\varkappa}\,\varphi^{\varkappa,-\ell-\varkappa-1}_{\tau',-\lambda'}(\omega)\,|\omega A_{12}+A_{22}|^{-2\ell-2\varkappa-2} \times$$
$$\times\ (\omega A_{12}+A_{22})^{\varkappa}\,\varphi^{\varkappa,\ell}_{\tau\lambda}(\omega\bar{A}) .$$

For calculating the integrals we introduce the points $\omega_{\pm}(A)$ defined by

$$(2.5.51)\quad \omega_{\pm}(A)\bar{A} = \pm 1 ,\ \text{i.e.}\ \ \omega_{\pm}(A) = (\pm A_{22}-A_{21})/(\mp A_{12}+A_{11})$$

on the boundary of the unit circle. With the aid of the relation

$$(2.5.52)\quad z\bar{A} - z'\bar{A} = \frac{z - z'}{(zA_{12}+A_{22})(z'A_{12}+A_{22})}$$

holding for $(z,z') \in \bar{\mathbb{C}}\times\bar{\mathbb{C}}$, $A \in SL(2,\mathbb{C})$, we can write for $V^{\varkappa,\ell}_{\tau'\lambda',\tau\lambda}(A)$

$$V^{\varkappa,\ell}_{\tau'\lambda',\tau\lambda}(A) = |A_{11}-A_{12}|^{-1-\ell-\varkappa/2-i\lambda}|A_{11}+A_{12}|^{-1-\ell-\varkappa/2+i\lambda} \times$$

(2.5.53)
$$\times \frac{1}{2\pi i}\int_{\partial K_{\tau'\tau}(A)} \frac{d\omega}{\omega}\,\varepsilon(\omega,A)^{\varkappa}\,|\omega-1|^{\ell+\varkappa/2+i\lambda'}\,|\omega+1|^{\ell+\varkappa/2-i\lambda'} \times$$

$$\times |\omega-\omega_+|^{-1-\ell-\varkappa/2-i\lambda}|\omega-\omega_-|^{-1-\ell-\varkappa/2+i\lambda} ,$$

$$\omega_\pm : = \omega_\pm(A) \;, \quad \varepsilon(\omega,A): = \omega^{-1/2}\frac{\omega A_{12}+A_{22}}{|\omega A_{12}+A_{22}|}\,\overline{\omega A}^{1/2} .$$

Since the functions $\varphi^{\varkappa,\ell}_{\tau\lambda}$ according to (2.5.4) are different from zero only on one half of the boundary of the unit circle, the integration path $\partial K_{\tau'\tau}(A)$ consists of the intersection of the arcs from τ' to $-\tau'$ and from ω_τ to $\omega_{-\tau}$ passed in the positive sense. For the further discussion we introduce the notation $\overline{\omega_1,\omega_2}$ for the arc between ω_1 and ω_2 on the boundary of the unit circle passed in the positive sense. With this we have

(2.5.54)
$$\partial K_{\tau'\tau}(A) = \overline{\tau',-\tau'} \cap \overline{\omega_\tau,\omega_{-\tau}} .$$

The somewhat tedious discussion of the sign factor $\varepsilon(\omega,A)$ is carried through in Appendix B with the result

(2.5.55)
$$\varepsilon(\omega,A) = \begin{cases} \text{sign Im}(A_{11}+A_{12}) & \text{for } \omega\in\overline{-1,\omega_-} , \\ -\text{sign Im}(A_{11}+A_{12}) & \text{for } \omega\in\overline{\omega_-,-1} . \end{cases}$$

The integral in (2.5.53) converges for $\omega_+ \neq \pm 1$, $\omega_- \neq \pm 1$ and $-1 < \text{Re}(\ell+\varkappa/2) < 0$. The last condition is fulfilled for principal and supplementary series. Because of the first conditions the matrix elements in the H_2-basis exist as functions in the ordinary sense only on the subset SU(1,1) - S with

(2.5.56)
$$S: = \bigcup_{\tau',\tau=\pm} S_{\tau'\tau} \,, \quad S_{\tau'\tau}: = \{A \in SU(1,1):\ \omega_{\tau'}(A) = \tau\} .$$

For the exact characterization of the singularity set $S \subset SU(1,1)$ we use the Iwasawa decomposition of SU(1,1) (cf. HELGASON [25], p. 234):

$$SU(1,1) = H_1\, N\, H_2' ,$$

(2.5.57)
$$H_2' = \left\{ \begin{pmatrix} \cosh\xi/2 & \sinh\xi/2 \\ \sinh\xi/2 & \cosh\xi/2 \end{pmatrix} : -\infty < \xi < \infty \right\} \simeq H_2/\{\mathbb{1}_2,-\mathbb{1}_2\} ,$$

$$N = \left\{ \begin{pmatrix} 1+i\frac{x}{2} & i\frac{x}{2} \\ -i\frac{x}{2} & 1-i\frac{x}{2} \end{pmatrix} : -\infty < x < \infty \right\},\ H_1 = \left\{ \begin{pmatrix} e^{i\varphi/2} & 0 \\ 0 & e^{-i\varphi/2} \end{pmatrix} : 0 \le \varphi < 4\pi \right\}.$$

With this one easily calculates the stability subgroup belonging to $\omega=1$:

$$\{A \in SU(1,1): \overline{1A^{-1}} = 1\} = N\, H_2 = S_{++}\,. \tag{2.5.58}$$

Using the element

$$\Gamma := i\sigma_3 = \begin{pmatrix} i & 0 \\ 0 & -i \end{pmatrix} \in SU(1,1) \tag{2.5.59}$$

which commutes the points ± 1 we get

$$S_{++} = N\, H_2,\ S_{+-} = \Gamma\, N\, H_2,\ S_{-+} = N\, H_2\, \Gamma^{-1},\ S_{--} = \Gamma\, N\, H_2\, \Gamma^{-1}. \tag{2.5.60}$$

Since S is a lower dimensional manifold in SU(1,1), the matrix elements in the H_2-basis represent ordinary functions for relative to the Haar measure on SU(1,1) almost all $A \in SU(1,1)$. For a more general definition of the matrix elements we would have to refer to the theory of distributions. So for instance for $A \in H_2 = S_{++} \cap S_{--}$ the matrix element relative to the H_2-basis is given by (2.5.49). The singularity manifold S divides SU(1,1) into six separate, connected pieces which are characterized by the following conditions:

$$\begin{aligned}
\text{I} &:= \{A \in SU(1,1):\ \mathrm{Im}\,\omega_+ > 0,\ \mathrm{Im}\,\omega_- < 0\}\,,\\
\text{I}' &:= \{A \in SU(1,1):\ \mathrm{Im}\,\omega_+ < 0,\ \mathrm{Im}\,\omega_- > 0\}\,,\\
\text{II} &:= \{A \in SU(1,1):\ \mathrm{Im}\,\omega_+ > 0,\ \mathrm{Im}\,\omega_- > 0,\ \mathrm{Re}\,\omega_+ > \mathrm{Re}\,\omega_-\}\,,\\
\text{II}' &:= \{A \in SU(1,1):\ \mathrm{Im}\,\omega_+ < 0,\ \mathrm{Im}\,\omega_- < 0,\ \mathrm{Re}\,\omega_+ > \mathrm{Re}\,\omega_-\}\,,\\
\text{III} &:= \{A \in SU(1,1):\ \mathrm{Im}\,\omega_+ > 0,\ \mathrm{Im}\,\omega_- > 0,\ \mathrm{Re}\,\omega_+ < \mathrm{Re}\,\omega_-\}\,,\\
\text{III}' &:= \{A \in SU(1,1):\ \mathrm{Im}\,\omega_+ < 0,\ \mathrm{Im}\,\omega_- < 0,\ \mathrm{Re}\,\omega_+ < \mathrm{Re}\,\omega_-\}\,.
\end{aligned} \tag{2.5.61}$$

The relations (B.13) and (B.18) of Appendix B enable us to define these pieces likewise by the conditions

$$\begin{aligned}
\text{I} &:= \{A \in SU(1,1):\ 0 < \mathrm{Im}_+\mathrm{Im}_- < 1,\ \mathrm{Re}_+\mathrm{Im}_+ < 0,\ \mathrm{Re}_-\mathrm{Im}_- < 0\}\,,\\
\text{I}' &:= \{A \in SU(1,1):\ 0 < \mathrm{Im}_+\mathrm{Im}_- < 1,\ \mathrm{Re}_+\mathrm{Im}_+ > 0,\ \mathrm{Re}_-\mathrm{Im}_- > 0\}\,,\\
\text{II} &:= \{A \in SU(1,1):\ \mathrm{Im}_+\mathrm{Im}_- < 0,\ \mathrm{Re}_+\mathrm{Im}_+ > 0,\ \mathrm{Re}_-\mathrm{Im}_- < 0\}\,,\\
\text{II}' &:= \{A \in SU(1,1):\ \mathrm{Im}_+\mathrm{Im}_- < 0,\ \mathrm{Re}_+\mathrm{Im}_+ < 0,\ \mathrm{Re}_-\mathrm{Im}_- > 0\}\,,
\end{aligned} \tag{2.5.62}$$

$$\text{III}: = \{A \in SU(1,1): \text{Im}_+\text{Im}_- > 1,\ \text{Re}_+\text{Im}_+ > 0,\ \text{Re}_-\text{Im}_- < 0\},$$

(2.5.62)
$$\text{III}': = \{A \in SU(1,1): \text{Im}_+\text{Im}_- > 1,\ \text{Re}_+\text{Im}_+ < 0,\ \text{Re}_-\text{Im}_- > 0\},$$

$$\text{Im}_\pm: = \text{Im}_\pm(A): = \text{Im}(A_{11} \pm A_{12}),\ \text{Re}_\pm: = \text{Re}_\pm(A): = \text{Re}(A_{11} \pm A_{12}).$$

We need not calculate the integrals for each of the six pieces separately, since all integrals can be obtained from those belonging to the pieces I and II with the aid of symmetry relations. With the element $\Gamma \in SU(1,1)$ defined in (2.5.59) we have

(2.5.63)
$$\omega_\pm(\Gamma A) = -\omega_\pm(A)\ ,\quad \omega_\pm(A\Gamma^{-1}) = \omega_\mp(A)\ .$$

The operation $A \longrightarrow \Gamma A$ commutes the regions I and I', II and III', III and II', the operation $A \longrightarrow A\Gamma^{-1}$ commutes the regions I and I', II and III, II' and III', while the composed operation $A \longrightarrow \Gamma A \Gamma^{-1}$ commutes the regions II and II', III and III' leaving invariant the regions I and I'. Thus by an operation of this kind we may carry over every $A \in SU(1,1) - S$ into one of the regions I or II. Since according to (B.15) and (B.16) the sign factor $\varepsilon(\omega,A)$ obeys the symmetry relations

(2.5.64)
$$\tau'\ \varepsilon(-\omega,\Gamma A) = \varepsilon(\omega,A) = \tau\ \varepsilon(\omega,A\Gamma^{-1})\ \text{for}\ \omega \in \partial K_{\tau'\tau}(A)\ ,$$

and since the integration path $\partial K_{\tau'\tau}(\Gamma A)$ is carried over to $\partial K_{-\tau',\tau}(A)$ by the substitution $\omega \longrightarrow -\omega$, while $\partial K_{\tau'\tau}(A\Gamma^{-1}) = \partial K_{\tau',-\tau}(A)$ holds, we get from (2.5.53) the following symmetry relations for $V^{\varkappa,\ell}_{\tau'\lambda',\tau\lambda}(A)$:

(2.5.65)
$$V^{\varkappa,\ell}_{\tau'\lambda',\tau\lambda}(A) = \tau'^{\varkappa}\ V^{\varkappa,\ell}_{-\tau',-\lambda',\tau\lambda}(\Gamma A) = \tau^{\varkappa}\ V^{\varkappa,\ell}_{\tau'\lambda',-\tau,-\lambda}(A\Gamma^{-1}) =$$
$$= (\tau'\tau)^{\varkappa}\ V^{\varkappa,\ell}_{-\tau',-\lambda',-\tau,-\lambda}(A^{-1\dagger})\ .$$

According to (2.5.50) this means for the matrix elements $U^{\varkappa,\ell,\eta}_{SU(1,1)}(A)_{\lambda'\lambda}$:

(2.5.66)
$$U^{\varkappa,\ell,\eta}_{SU(1,1)}(A)_{\lambda'\lambda} = \Gamma^{\varkappa,\ell}(\lambda')^{-1}\ U^{\varkappa,\ell,\eta}_{SU(1,1)}(\Gamma A)_{-\lambda',\lambda} =$$
$$= U^{\varkappa,\ell,\eta}_{SU(1,1)}(A\Gamma^{-1})_{\lambda',-\lambda}\ \Gamma^{\varkappa,\ell}(\lambda) =$$
$$= \Gamma^{\varkappa,\ell}(\lambda')^{-1}\ U^{\varkappa,\ell,\eta}_{SU(1,1)}(A^{-1\dagger})_{-\lambda',-\lambda}\ \Gamma^{\varkappa,\ell}(\lambda)\ ,$$

$$\Gamma^{\varkappa,\ell}(\lambda) := N^{\varkappa,\ell}(-\lambda)^{-1}\,\sigma_1(\sigma_3)^{\varkappa}\,N^{\varkappa,\ell}(\lambda) = \tag{2.5.66}$$

$$= -\frac{\Gamma(-\ell-\varkappa/2+i\lambda)\Gamma(1+\ell+\varkappa/2+i\lambda)}{\pi}\, i^{\varkappa}\begin{pmatrix}\sin\pi(i\lambda-\frac{\varkappa}{2}) & (-)^{\varkappa}\sin\pi\ell\\ \sin\pi\ell & \sin\pi(i\lambda+\frac{\varkappa}{2})\end{pmatrix}.$$

Obviously we have

$$U^{\varkappa,\ell,\eta}_{SU(1,1)}(\Gamma)_{\lambda'\lambda} = \Gamma^{\varkappa,\ell}(\lambda)\,\delta(\lambda'+\lambda) \quad . \tag{2.5.67}$$

These relations permit the calculation of the matrix elements for each of the six pieces if they are known in the regions I and II. We finally complete (2.5.66) by a somewhat more convenient symmetry relation. Because of

$$\omega_{\pm}(A^{*}) = 1/\omega_{\pm}(A) \tag{2.5.68}$$

the operation $A \longrightarrow A^{*} = \sigma_1 A \sigma_1$ commutes the regions I and I', II and II', III and III'. Since according to (B.21) we have

$$\varepsilon(\omega,A^{*}) = \varepsilon(1/\omega,A) \ , \tag{2.5.69}$$

and since $\partial K_{\tau'\tau}(A^{*})$ is carried over to $\partial K_{-\tau',-\tau}(A)$ by the substitution $\omega \longrightarrow 1/\omega$, from (2.5.53) we get the relation

$$V^{\varkappa,\ell}_{\tau'\lambda',\tau\lambda}(A^{*}) = V^{\varkappa,\ell}_{-\tau',\lambda',-\tau,\lambda}(A) \ . \tag{2.5.70}$$

Since for $N^{\varkappa,\ell}(\lambda)$ in (2.5.35) we have

$$\sigma_1\, N^{\varkappa,\ell}(\lambda)\, \sigma_1 = N^{\varkappa,\ell}(\lambda) \tag{2.5.71}$$

we get from (2.5.50) the symmetry relation

$$U^{\varkappa,\ell,\eta}_{SU(1,1)}(A)_{\lambda'\lambda} = \sigma_1\, U^{\varkappa,\ell,\eta}_{SU(1,1)}(A^{*})_{\lambda'\lambda}\,\sigma_1 \ . \tag{2.5.72}$$

Analogs of the middle relation in (2.5.66) cannot be set up in this case because the operation $A \longrightarrow A^{*} = \sigma_1 A \sigma_1$, contrary to the operation $A \longrightarrow A^{-1\dagger} = \Gamma A \Gamma^{-1}$, cannot be effected by an inner automorphism in SU(1,1).

With the aid of the integral formula

$$\frac{1}{2\pi i}\int_{\omega_1}^{\omega_2}\frac{d\omega}{\omega}|\omega-\omega_1|^{a_1}|\omega-\omega_2|^{a_2}|\omega-\omega_3|^{a_3}|\omega-\omega_4|^{a_4} = \frac{\Gamma(a_1+1)\Gamma(a_2+1)}{2\pi\,\Gamma(a_1+a_2+2)}\times$$

(2.5.73) $$|\omega_2-\omega_1|^{a_1+a_2+1}|\omega_3-\omega_1|^{a_1+a_3+1}|\omega_3-\omega_2|^{-a_1-1}|\omega_4-\omega_1|^{a_4}\times$$

$$\times F(-a_4,a_1+1;a_1+a_2+2;\frac{(\omega_4-\omega_3)(\omega_2-\omega_1)}{(\omega_4-\omega_1)(\omega_2-\omega_3)})\ ,\quad a_1+a_2+a_3+a_4+2 = 0\ ,$$

the matrix elements $V^{\varkappa,\ell}_{\tau'\lambda',\tau\lambda}(A)$ may be expressed by the hypergeometric function $F \equiv {}_2F_1$. Using functional equations for the hypergeometric and Γ-functions we can summarize the results for $A\in I$ and $A\in II$, respectively, in the formula

$$U^{\varkappa,\ell,\eta}_{SU(1,1)}(A)_{\tau'\lambda',\tau\lambda} = (-\varepsilon\,\mathrm{sign}\ Im_+)^{\varkappa}\Big\{\frac{\delta_{\tau'\tau}}{2\sin\pi i(\lambda'-\lambda)}\times$$

$$\times\Big[a^{\varkappa,\ell}_{\lambda'\lambda}e^{\varepsilon\tau\pi(\lambda'-\lambda)/2}F^{\varkappa,\ell}_{\lambda'\lambda}(A) - e^{-\tau\pi(\lambda'-\lambda)/2}F^{\varkappa,\ell}_{-\lambda',-\lambda}(A^{-1\dagger})\Big] +$$

$$+\varepsilon(-)^{\varkappa}\frac{\sin\pi\ell}{2\pi^2}b^{\varkappa,\ell}_{\lambda'\lambda}F^{\varkappa,\ell}_{\lambda'\lambda}(A)\times$$

$$\times\Big[i\tau\delta_{\tau'\tau}\sin\pi\ell\, e^{\varepsilon\tau\pi(\lambda'-\lambda)/2} - \varepsilon(i\varepsilon\tau)^{\varkappa}\delta_{\tau',-\tau}e^{-\varepsilon\tau\pi(\lambda'+\lambda)/2}\Big]\Big\}\ ,$$

$\varepsilon = +1$ for $A\in I$, $\varepsilon = -1$ for $A\in II$,

(2.5.74)

$$2a^{\varkappa,\ell}_{\lambda'\lambda} := \frac{\Gamma(1+\ell+\varkappa/2-i\lambda')\Gamma(1+\ell+\varkappa/2+i\lambda)}{\Gamma(1+\ell+\varkappa/2+i\lambda')\Gamma(1+\ell+\varkappa/2-i\lambda)}+\frac{\Gamma(-\ell-\varkappa/2-i\lambda')\Gamma(-\ell-\varkappa/2+i\lambda)}{\Gamma(-\ell-\varkappa/2+i\lambda')\Gamma(-\ell-\varkappa/2-i\lambda)}\ ,$$

$$b^{\varkappa,\ell}_{\lambda'\lambda} := \Gamma(1+\ell+\varkappa/2-i\lambda')\Gamma(1+\ell+\varkappa/2+i\lambda)\Gamma(-\ell-\varkappa/2-i\lambda')\Gamma(-\ell-\varkappa/2+i\lambda)\ ,$$

$$F^{\varkappa,\ell}_{\lambda'\lambda}(A) := |Re_-Im_+|^{-i\lambda'}|Re_-/Im_+|^{-i\lambda}\frac{1}{\Gamma(1-i(\lambda'-\lambda))}\times$$

$$\times F(-\ell-\varkappa/2-i\lambda',1+\ell+\varkappa/2-i\lambda';1-i(\lambda'-\lambda);Im_+Im_-)\ ,$$

$$Im_\pm := Im(A_{11}\pm A_{12})\ ,\quad Re_\pm := Re(A_{11}\pm A_{12})\ .$$

For the discrete series we get the matrix elements by the following consideration: One can continue analytically the integrals (2.5.53) for $V^{\varkappa,\ell}_{\lambda'\lambda}(A)$ in ℓ beyond the strip $-1 < Re(\ell+\varkappa/2) < 0$ and show that according to (2.5.50) by continuation to integer values of $\ell\geq 0$ they yield the matrix elements of the discrete series. Since the explicit expressions (2.5.74) are already continuable in ℓ, they can be accepted for the discrete series simply by setting $\ell\geq 0$, integer. Because of the factor $\sin\pi\ell$ the second bracket containing the non-diagonal part in τ',τ is not present, whereby the decay of the representations on $\tilde{\mathfrak{h}}$ into the direct sum of the parts belonging to $\eta = +$ and $\eta = -$ is demonstrated explicitly once more. Furthermore we point out that the matrix $\Gamma^{\varkappa,\ell}(\lambda)$ in (2.5.66) becomes diagonal also for the discrete series, so that the symmetry relations (2.5.66) can be formulated for both parts of this series separately. The symmetry relation (2.5.72), however, reflects the antiunitary equivalence of the representations $U^{\varkappa,\ell,+}_{SU(1,1)}$ and $U^{\varkappa,\ell,-}_{SU(1,1)}$ known from (1.3.51).

In Sections 2.2 - 2.4 we have solved the reduction problem for the restriction to the compact subgroup H_1 of the representation U^{ρ}_G by presenting a suitable orthonormal basis of the representation space $\mathfrak{h}^{\rho}_G$ in which $U^{\rho}_G|H_1$ decayed into a direct sum. We now may interpret these basis elements as kernels of a unitary map of the representation space $\mathfrak{h}^{\rho}_G$ onto the Hilbert space $\tilde{\mathfrak{h}}^{\rho}_{G,H_1}$ of complex valued sequences, defined on the set of the basis elements with finite sum of absolute squares, on which $U^{\rho}_G|H_1$ decays into a direct sum. This remark permits us to summarize formally the results for the compact group H_1 and the noncompact group H_2 in the following way: Let $\hat{H}$ be the set of all equivalence classes of irreducible unitary representations of H, $H \in \{H_1, H_2\}$, $\sigma \in \hat{H}$, $\hat{H}_{\rho}$ the set of all $\sigma \in \hat{H}$ which occur in the reduction of $U^{\rho}_G|H$, χ^{σ} an irreducible unitary representation of H, $n(\rho,\sigma) \equiv n(\rho)$ the multiplicity of χ^{σ} in $U^{\rho}_G|H$, $\hat{\nu}$ the Plancherel measure on $\hat{H}$. Then in any case we have constructed an equivalence transformation $\breve{A}$ which maps $\mathfrak{h}^{\rho}_G$ unitarily onto the Hilbert space

$$\breve{A}\,\mathfrak{h}^{\rho}_G = \tilde{\mathfrak{h}}^{\rho}_{G,H} = \oplus\int_{\hat{H}_{\rho}} \sqrt{d\hat{\nu}(\sigma)}\; \mathbb{C}^{n(\rho)} \; . \tag{2.5.75}$$

$\tilde{\mathfrak{h}}^{\rho}_{G,H}$ consists of the complex valued functions $\tilde{f} = \breve{A}f$, $f \in \mathfrak{h}^{\rho}_G$, on $\{1,2, \dots ,n(\rho)\} \times \hat{H}_{\rho}$ with scalar product

$$\langle \tilde{f} | \tilde{g} \rangle^{\sim\rho}_{G,H} := \int_{\hat{H}_{\rho}} d\hat{\nu}(\sigma) \sum_{\tau=1}^{n(\rho)} \tilde{f}_{\tau}(\sigma)^{*}\, \tilde{g}_{\tau}(\sigma) = \langle f|g \rangle^{\rho}_G \; . \tag{2.5.76}$$

For the representation $\tilde{U}^{\rho}_G = \breve{A}\, U^{\rho}_G\, \breve{A}^{-1}$ on $\tilde{\mathfrak{h}}^{\rho}_{G,H}$ we have

$$\begin{aligned} \tilde{U}^{\rho}_G(Q)\, \tilde{f}_{\tau}(\sigma) &= \chi^{\sigma}(Q)\, \tilde{f}_{\tau}(\sigma) \;, \quad Q \in H \;, \\ \tilde{U}^{\rho}_G | H &= \oplus\int_{\hat{H}_{\rho}} d\hat{\nu}(\sigma) \Big[\oplus \sum_{1}^{n(\rho)} \chi^{\sigma} \Big] \; . \end{aligned} \tag{2.5.77}$$

2.6 Orthogonality and Completeness Relations for the Irreducible Unitary Representations of SU(2), SU(1,1) and E(2)

We start from the complex valued function on the complex ℓ- and z-planes

$$(2.6.1)\quad \begin{aligned} u^{\varkappa,\ell}_{\mu'\mu}(z) &:= N^{\varkappa,\ell}_{\mu'\mu}(-z)^{(\mu'-\mu)/2}(1-z)^{(\mu'+\mu+\varkappa)/2}\,\frac{F(\mu'-\ell,\mu'+\ell+\varkappa+1;\mu'-\mu+1;z)}{\Gamma(\mu'-\mu+1)}\,,\\ N^{\varkappa,\ell}_{\mu'\mu} &:= (-)^{\mu'-\mu}\,\frac{N^{\varkappa,\ell}_{\mu}\,\Gamma(\mu'+\ell+\varkappa+1)}{N^{\varkappa,\ell}_{\mu'}\,\Gamma(\mu+\ell+\varkappa+1)} = (-)^{\mu'-\mu}\left[\frac{\Gamma(\mu'+\ell+\varkappa+1)\Gamma(\mu'-\ell)}{\Gamma(\mu+\ell+\varkappa+1)\Gamma(\mu-\ell)}\right]^{1/2}\,,\end{aligned}$$

with $\varkappa\in\{0,1\}$ and integer μ',μ . $N^{\varkappa,\ell}_{\mu}$ is defined in (2.3.4). The hypergeometric function $F\equiv {}_2F_1$ is an entire function in the ℓ-plane and a single-valued analytical function in the z-plane cut along the real axis from 1 to $+\infty$ with the value 1 at $z=0$. Further cuts from 0 to $+\infty$ and from 1 to $+\infty$ in the z-plane, arising from the factor $(-z)^{(\mu'-\mu)/2}(1-z)^{(\mu'+\mu+\varkappa)/2}$, must be added in dependence on the values of the exponents. $N^{\varkappa,\ell}_{\mu'\mu}$ is a single-valued analytical function in the ℓ-plane cut along finitely many intervals with integer endpoints on the real axis with the value

$$(2.6.2)\quad N^{\varkappa,\ell\pm io}_{\mu'\mu} = e^{\pm i\pi\frac{\mu'-\mu}{2}}\sqrt[+]{\frac{\Gamma(\mu'+\ell+\varkappa+1)\Gamma(-\mu+\ell+1)}{\Gamma(\mu+\ell+\varkappa+1)\Gamma(-\mu'+\ell+1)}}\quad\text{for } \ell\geq\max(\mu',\mu,-\mu'-\varkappa,-\mu-\varkappa)\,,$$

which is derived from the behaviour of the function $N^{\varkappa,\ell}_{\mu}$ in the complex ℓ-plane as discussed following (2.3.4). From the symmetry relations (2.3.6) for $N^{\varkappa,\ell}_{\mu}$ we have

$$(2.6.3)\quad N^{\varkappa,\ell}_{\mu'\mu} = 1/N^{\varkappa,\ell}_{\mu\mu'} = N^{\varkappa,\ell}_{-\mu-\varkappa,-\mu'-\varkappa} = N^{\varkappa,-\ell-\varkappa-1}_{\mu'\mu}\,,$$

and therefore with functional relations between the hypergeometric functions ([23] , p. 105) we get

$$(2.6.4)\quad \begin{aligned} (-)^{\mu'-\mu}\,u^{\varkappa,\ell}_{-\mu'-\varkappa,-\mu-\varkappa}(z) &= u^{\varkappa,\ell}_{\mu'\mu}(z) = u^{\varkappa,\ell}_{-\mu-\varkappa,-\mu'-\varkappa}(z)\,,\\ u^{\varkappa,-\ell-\varkappa-1}_{\mu'\mu}(z) &= u^{\varkappa,\ell}_{\mu'\mu}(z)\,.\end{aligned}$$

So for the following discussion of $u^{\varkappa,\ell}_{\mu'\mu}$ we may presume

$$(2.6.5)\quad -\mu'-\varkappa\leq\mu\leq\mu',\ \mu'\geq 0\,.$$

On the negative real axis of the z-plane and for the ℓ-values of the irreducible unitary representations of SU(1,1) $u^{\varkappa,\ell}_{\mu'\mu}$, according to (2.3.15), is identical with $\hat{u}^{\varkappa,\ell}_{\mu'\mu}$:

$$u^{\varkappa,\ell}_{\mu'\mu}(x) = \hat{u}^{\varkappa,\ell}_{\mu'\mu}(x) \ , \quad x \leq 0 \ . \tag{2.6.6}$$

The functions $\tilde{u}^{\varkappa,\ell}_{\mu'\mu}$ and $\tilde{u}'^{\varkappa,\ell}_{\mu'\mu}$ defined by

$$\begin{aligned} u^{\varkappa,\ell}_{\mu'\mu}(z) &= :e^{i\pi\,\mathrm{sign}(\mathrm{Im}\,\ell)(\mu'-\mu)/2}\,\tilde{u}^{\varkappa,\ell}_{\mu'\mu}(z) = \\ &= :e^{i\pi(\mathrm{sign}(\mathrm{Im}\,\ell)\,-\,\mathrm{sign}(\mathrm{Im}\,z))(\mu'-\mu)/2}\,\tilde{u}'^{\varkappa,\ell}_{\mu'\mu}(z) \end{aligned} \tag{2.6.7}$$

according to (2.6.2) for the integer ℓ-values of the irreducible unitary representations of SU(2) coincide with the corresponding functions from (2.2.6) and (2.2.7). The function

$$U^{\varkappa,\ell}(A)_{\mu'\mu} : = (A_{11}/A_{22})^{(\mu'+\mu+\varkappa)/2}(A_{12}/A_{21})^{(\mu'-\mu)/2}\, u^{\varkappa,\ell}_{\mu'\mu}(-A_{12}A_{21}) \tag{2.6.8}$$

on SL(2,$\mathbb{C}$) therefore interpolates analytically between the matrix elements $U^{\varkappa,\ell}_{SU(2)}(A)_{\mu'\mu}$ and $U^{\varkappa,\ell,\eta}_{SU(1,1)}(A)_{\mu'\mu}$. For by restriction to SU(2) according to (2.6.7) and (2.2.5) we get for integer $\ell \geq \mu'$

$$U^{\varkappa,\ell\pm io}(A)_{\mu'\mu} = e^{\pm i\pi(\mu'-\mu)/2}\, U^{\varkappa,\ell}_{SU(2)}(A) \ , \quad A \in SU(2) \ , \tag{2.6.9}$$

and by restriction to SU(1,1) according to (2.6.6) and (2.3.15) we get for the ℓ-values of the series of irreducible unitary representations of SU(1,1)

$$U^{\varkappa,\ell}(A)_{\mu'\mu} = U^{\varkappa,\ell,\eta}_{SU(1,1)}(A)_{\mu'\mu} \ , \quad A \in SU(1,1) \ . \tag{2.6.10}$$

The parametrization by Eulers angles yields a special form of this analytical interpolation. If we set

$$\begin{aligned} A = A(\alpha,\zeta,\gamma) : &= \begin{pmatrix} e^{i\alpha/2} & 0 \\ 0 & e^{-i\alpha/2} \end{pmatrix} \begin{pmatrix} \cosh\zeta/2 & \sinh\zeta/2 \\ \sinh\zeta/2 & \cosh\zeta/2 \end{pmatrix} \begin{pmatrix} e^{i\gamma/2} & 0 \\ 0 & e^{-i\gamma/2} \end{pmatrix}, \\ &0 \leq \alpha < 2\pi \ , \ 0 \leq \gamma < 4\pi \ , \\ \zeta \in S_{-\pi,\pi} : &= \{\zeta = \xi + i\beta : 0 < \xi < \infty, \ -\pi < \beta \leq \pi\} \cup \{\zeta = i\beta : 0 \leq \beta \leq \pi\} \ , \end{aligned} \tag{2.6.11}$$

we get a parametrization of SU(2) for $\xi := \mathrm{Re}\,\zeta = 0$ and a parametrization of SU(1,1) for $\beta := \mathrm{Im}\,\zeta = 0$. According to (2.6.8) we have

$$(2.6.12)\quad U^{\varkappa,\ell}(A(\alpha,\zeta,\gamma))_{\mu'\mu} = e^{i\alpha(\mu'+\varkappa/2)}\, u^{\varkappa,\ell}_{\mu'\mu}(-\sinh^2\zeta/2)\, e^{i\gamma(\mu+\varkappa/2)} .$$

The analytical continuation from SU(2) to SU(1,1) in this case takes place within the strip $S_{-\pi,\pi}$ which by

$$(2.6.13)\qquad z = -\sinh^2\zeta/2$$

is mapped onto the complex z-plane such that the sections ξ = const and the half lines β = const are mapped onto the ellipses with focal points 0 and 1 and the half branches of the hyperbolas orthogonal to these, starting from the interval (0,1), respectively. Here $z = x - io$, $x \geqslant 1$ is the image of $\beta = \pi - o$. To the interval $[0,1]$ and the negative real axis in the z-plane there correspond the straight line pieces $\xi = 0$ and $\beta = 0$ from $S_{-\pi,\pi}$, respectively.

The function $u^{\varkappa,\ell}_{\mu'\mu}$ in (2.6.1) can be expressed with the aid of the Jacobi function $P_n^{(\alpha,\beta)}$ ([26], p. 170) in the following way:

$$(2.6.14)\quad u^{\varkappa,\ell}_{\mu'\mu}(z) = \frac{N^{\varkappa,\ell}_{\mu'}}{N^{\varkappa,\ell}_{\mu}}(-z)^{(\mu'-\mu)/2}(1-z)^{(\mu'+\mu+\varkappa)/2}\, P^{(\mu'-\mu,\mu'+\mu+\varkappa)}_{\ell-\mu'}(1-2z) .$$

We use this formula to define, with the aid of the Jacobi functions of the second kind $Q_n^{(\alpha,\beta)}$ ([26], p. 170), the functions

$$v^{\varkappa,\ell}_{\mu'\mu}(z) := \frac{N^{\varkappa,\ell}_{\mu'}}{N^{\varkappa,\ell}_{\mu}}(-z)^{(\mu'-\mu)/2}(1-z)^{(\mu'+\mu+\varkappa)/2}\, Q^{(\mu'-\mu,\mu'+\mu+\varkappa)}_{\ell-\mu'}(1-2z) =$$

$$(2.6.15)\qquad = \frac{1}{2}\Gamma(1+\ell-\mu')\Gamma(1+\ell+\mu+\varkappa)\, N^{\varkappa,\ell}_{\mu'\mu}\,(-z)^{(\mu'-\mu)/2}(1-z)^{(\mu'+\mu+\varkappa)/2}\times$$

$$\times(-z)^{-\mu'-\ell-\varkappa-1}\,\frac{F(\mu'+\ell+\varkappa+1,\mu+\ell+\varkappa+1;2\ell+\varkappa+2;1/z)}{\Gamma(2\ell+\varkappa+2)} .$$

In the z-plane $v^{\varkappa,\ell}_{\mu'\mu}$ has a cut between 0 and 1 on the real axis stemming from F, a logarithmic cut along the positive real axis from the function $(-z)^{-\mu'-\ell-\varkappa-1}$ for noninteger ℓ, as well as root cuts from 0 to $+\infty$ and from 1 to $+\infty$, respectively from the factor $(-z)^{(\mu'-\mu)/2}(1-z)^{(\mu'+\mu+\varkappa)/2}$ in dependence on the values of the exponents. In the ℓ-plane $v^{\varkappa,\ell}_{\mu'\mu}$, in addition to the cuts of $N^{\varkappa,\ell}_{\mu'\mu}$, has the poles stemming from the Γ-functions in the numerator, while the rest is an entire function in ℓ. For $\mu' = \mu = \varkappa = 0$ we have

$$u_{00}^{0,\ell}(z) = P_\ell(1-2z)\ , \qquad v_{00}^{0,\ell}(z) = Q_\ell(1-2z) \tag{2.6.16}$$

with the Legendre functions P_ℓ and Q_ℓ of the first and second kind. Our next aim is the derivation of a pendant for the functions $u_{\mu'\mu}^{\varkappa,\ell}$ and $v_{\mu'\mu}^{\varkappa,\ell}$ to the Heine formula ([23], p. 168)

$$2\sum_{n=0}^{\infty}(2n+1)\,P_n(1-2z)\,Q_n(1-2z') = \frac{1}{z-z'} \tag{2.6.17}$$

which holds whenever z and z' are separated by an ellipse with focal points 0 and 1 with z in the interior and z' in the exterior. To this end we first discuss some properties of the functions $u_{\mu'\mu}^{\varkappa,\ell}$ and $v_{\mu'\mu}^{\varkappa,\ell}$.

We start from the formula

$$\begin{aligned}(-)^{\mu'-\mu}\frac{2\sin\pi\ell}{\pi}\,v_{\mu'\mu}^{\varkappa,\ell}(z) &= e^{i\pi\ell\,\mathrm{sign}(\mathrm{Im}\,z)}\,u_{\mu'\mu}^{\varkappa,\ell}(z) - \\ &- e^{i\pi(\mu+\varkappa/2)\mathrm{sign}(\mathrm{Im}\,\ell)}\,e^{-i\pi\varkappa\,\mathrm{sign}(\mathrm{Im}\,z)/2}\,u_{\mu',-\mu-\varkappa}^{\varkappa,\ell}(1-z)\end{aligned} \tag{2.6.18}$$

which is equivalent to the relation between the corresponding hypergeometric functions ([23], p. 106). From this and the symmetry relations (2.6.4) for the $u_{\mu'\mu}^{\varkappa,\ell}$ we get the corresponding relations for the $v_{\mu'\mu}^{\varkappa,\ell}$:

$$(-)^{\mu'-\mu}\,v_{-\mu'-\varkappa,-\mu-\varkappa}^{\varkappa,\ell}(z) = v_{\mu'\mu}^{\varkappa,\ell}(z) = v_{-\mu-\varkappa,-\mu'-\varkappa}^{\varkappa,\ell}(z)\ . \tag{2.6.19}$$

So we may assume the validity of (2.6.5). From the Gaussian recursion relations for "neighboured" hypergeometric functions ([23], p. 103) one derives the recursion formula

$$\begin{aligned}\frac{1}{2}\left[1-2z-\frac{(2\mu'+\varkappa)(2\mu+\varkappa)}{(2\ell+\varkappa)(2\ell+\varkappa+2)}\right]u_{\mu'\mu}^{\varkappa,\ell}(z) &= \\ = \frac{[(\ell-\mu')(\ell+\mu'+\varkappa)(\ell-\mu)(\ell+\mu+\varkappa)]^{1/2}}{(2\ell+\varkappa)(2\ell+\varkappa+1)}\,u_{\mu'\mu}^{\varkappa,\ell-1}(z) &+ \\ + \frac{[(1+\ell-\mu')(1+\ell+\mu'+\varkappa)(1+\ell-\mu)(1+\ell+\mu+\varkappa)]^{1/2}}{(2\ell+\varkappa+1)(2\ell+\varkappa+2)}\,u_{\mu'\mu}^{\varkappa,\ell+1}(z)\end{aligned} \tag{2.6.20}$$

with positive values of the roots for $\ell>\mu'$. According to (2.6.18) the same formula holds for the functions $v_{\mu'\mu}^{\varkappa,\ell}$:

$$\frac{1}{2}\left[1 - 2z - \frac{(2\mu'+\varkappa)(2\mu+\varkappa)}{(2\ell+\varkappa)(2\ell+\varkappa+2)}\right] v^{\varkappa,\ell}_{\mu'\mu}(z) =$$

$$(2.6.21) \qquad = \frac{[(\ell-\mu')(\ell+\mu'+\varkappa)(\ell-\mu)(\ell+\mu+\varkappa)]^{1/2}}{(2\ell+\varkappa)(2\ell+\varkappa+1)} v^{\varkappa,\ell-1}_{\mu'\mu}(z) +$$

$$+ \frac{[(1+\ell-\mu')(1+\ell+\mu'+\varkappa)(1+\ell-\mu)(1+\ell+\mu+\varkappa)]^{1/2}}{(2\ell+\varkappa+1)(2\ell+\varkappa+2)} v^{\varkappa,\ell+1}_{\mu'\mu}(z) .$$

A simple application of both recursion relations yields

$$2(2\ell+\varkappa+1)(z-z')u^{\varkappa,\ell}_{\mu'\mu}(z)\, v^{\varkappa,\ell}_{\mu'\mu}(z') = K^{\varkappa,\ell-1}_{\mu'\mu}(z,z')-K^{\varkappa,\ell}_{\mu'\mu}(z,z') ,$$

$$(2.6.22) \quad K^{\varkappa,\ell-1}_{\mu'\mu}(z,z') := \frac{[(\ell-\mu')(\ell+\mu'+\varkappa)(\ell-\mu)(\ell+\mu+\varkappa)]^{1/2}}{\ell+\varkappa/2} \times$$

$$\times\left[u^{\varkappa,\ell}_{\mu'\mu}(z)\, v^{\varkappa,\ell-1}_{\mu'\mu}(z') - u^{\varkappa,\ell-1}_{\mu'\mu}(z)\, v^{\varkappa,\ell}_{\mu'\mu}(z')\right] .$$

If we sum up this formula over a series ℓ_o, ℓ_o+1, ... , ℓ_o+N of ℓ-values with integer distances, we get

$$(2.6.23) \quad 2(z-z')\sum_{\ell=\ell_o}^{\ell_o+N}(2\ell+\varkappa+1)u^{\varkappa,\ell}_{\mu'\mu}(z)v^{\varkappa,\ell}_{\mu'\mu}(z') = K^{\varkappa,\ell_o-1}_{\mu'\mu}(z,z')-K^{\varkappa,\ell_o+N}_{\mu'\mu}(z,z').$$

From the relation $F(a,b;a;z) = (1-z)^{-b}$ follows

$$(2.6.24) \quad \lim_{\ell\to\mu'\pm io} \frac{[(\ell-\mu')(\ell+\mu'+\varkappa)(\ell-\mu)(\ell+\mu+\varkappa)]^{1/2}}{\ell+\varkappa/2} v^{\varkappa,\ell-1}_{\mu'\mu}(z) = \frac{1}{u^{\varkappa,\mu'\mp io}_{\mu'\mu}(z)} ,$$

$$u^{\varkappa,\mu'\mp io}_{\mu'\mu}(z) = e^{\mp i\pi\frac{\mu'-\mu}{2}}\,{}_{+}\!\sqrt{\frac{(2\mu'+\varkappa)!}{(\mu'-\mu)!\,(\mu'+\mu+\varkappa)!}}(-z)^{(\mu'-\mu)/2}(1-z)^{(\mu'+\mu+\varkappa)/2}.$$

With $\lim_{\ell\to\mu'-1\pm io} u^{\varkappa,\ell}_{\mu'\mu}(z) = 0$ we get

$$(2.6.25) \quad \lim_{\ell\to\mu'\pm io} K^{\varkappa,\ell-1}_{\mu'\mu}(z,z') = (-)^{\mu'-\mu}\, \omega^{\varkappa}_{\mu'\mu}(z)/\,\omega^{\varkappa}_{\mu'\mu}(z') ,$$

$$\omega^{\varkappa}_{\mu'\mu}(z) := (-z)^{(\mu'-\mu)/2}(1-z)^{(\mu'+\mu+\varkappa)/2}$$

With the aid of the asymptotic expansions for $|\ell| \longrightarrow \infty$, $|\mathrm{arc}\,\ell|<\pi$,

$$u^{\varkappa,\ell}_{\mu'\mu}(-\sinh^2\varsigma/2) = \frac{e^{i\pi\,\mathrm{sign}(\mathrm{Im}\ell)(\mu'-\mu)/2}}{(2\pi\ell\sinh\varsigma)^{1/2}}\left[e^{\varsigma(\ell+1/2+\varkappa/2)} +\right.$$

$$(2.6.26) \qquad \left. + e^{i\pi\,\mathrm{sign}(\mathrm{Im}\varsigma)(\mu'-\mu+1/2)}\, e^{-\varsigma(\ell+1/2+\varkappa/2)}\right] ,$$

$$v^{\varkappa,\ell}_{\mu'\mu}(-\sinh^2\varsigma/2) = \frac{e^{i\pi\,\mathrm{sign}(\mathrm{Im}\ell)(\mu'-\mu)/2}}{(2\pi\ell\sinh\varsigma)^{1/2}}\, e^{-\varsigma(\ell+1/2+\varkappa/2)}, \; |\mathrm{arc}\,\ell|<\pi,$$

which follow from the corresponding Watson expansions of hypergeometric functions ([23], p. 77) together with the lower symmetry relation in (2.6.4) we get

$$(2.6.27)\quad \begin{aligned}&\lim_{\mathrm{Re}\,\ell\to+\infty} K^{\varkappa,\ell}_{\mu'\mu}(-\sinh^2\varsigma/2,-\sinh^2\varsigma'/2) = \\ &= \frac{(-)^{\mu'-\mu}(e^{\varsigma'}-e^{\varsigma})}{2(\sinh\varsigma'\sinh\varsigma)^{1/2}}\lim_{\mathrm{Re}\,\ell\to+\infty} e^{-(\varsigma'-\varsigma)(\ell+1/2+\varkappa/2)} = 0 \text{ for } \varsigma'>\varsigma\,,\end{aligned}$$

i.e. $K^{\varkappa,\ell}_{\mu'\mu}(z,z')$ vanishes for $\mathrm{Re}\,\ell \longrightarrow +\infty$ if z lies in the interior E(z') of the ellipse through z' with focal points 0 and 1. If we take into account that the product $u^{\varkappa,\ell}_{\mu'\mu}(z)\; v^{\varkappa,\ell}_{\mu'\mu}(z')$ has no cuts in the ℓ-plane, we get from (2.6.23) with (2.6.25) and (2.6.27)

$$(2.6.28)\quad \sum_{\ell=\mu'}^{\infty}(2\ell+\varkappa+1)\; u^{\varkappa,\ell}_{\mu'\mu}(z)\; v^{\varkappa,\ell}_{\mu'\mu}(z') = \frac{(-)^{\mu'-\mu}\omega^{\varkappa}_{\mu'\mu}(z)}{2\,\omega^{\varkappa}_{\mu'\mu}(z')}\,\frac{1}{z-z'},\quad z\in E(z').$$

Here the series on the left hand side converges for all $z\in E(z')$ and represents there the function standing on the right hand side. Obviously (2.6.28) is a generalization of the Heine formula (2.6.17) the latter being obtained for $\mu' = \mu = \varkappa = 0$.

The formula (2.6.28) is adapted for the application to SU(2)-representations. For the SU(1,1)-representations we need a modification of this formula which we shall derive with the aid of a sort of Sommerfeld-Watson transformation ([27], p. 203 ff.). To this end we first represent the sum in (2.6.28) as an integral over the loop C_1

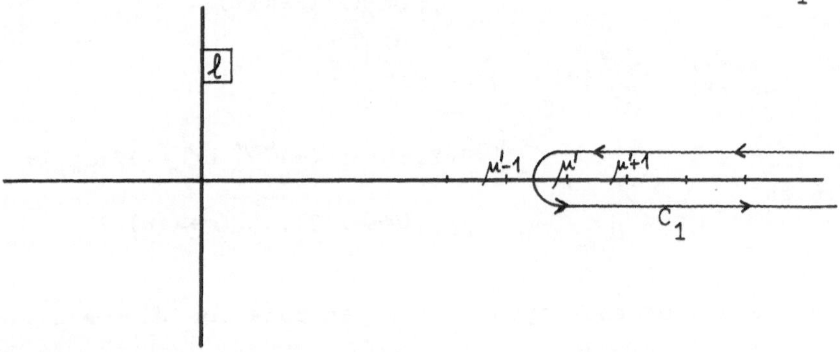

Figure 2.1

as sketched in Figure 2.1:

$$(2.6.29)\quad -\frac{i}{2}\int_{C_1}\frac{d\ell}{\sin\pi\ell}(2\ell+\varkappa+1)\; u^{\varkappa,\ell}_{\mu'\mu}(z)\; e^{-i\pi\ell\,\mathrm{sign}(\mathrm{Im}\,z')}\; v^{\varkappa,\ell}_{\mu'\mu}(z')\;.$$

Since the function $u^{\varkappa,\ell}_{\mu'\mu}(z)\; v^{\varkappa,\ell}_{\mu'\mu}(z')$ according to (2.6.2) has no singularities or cuts for $\mathrm{Re}\,\ell>\mu'-1$ by an application of the residue theorem we get from (2.6.29) the sum (2.6.28). Here the factor $(-)^{\ell}$ for integer ℓ was given the somewhat exotic form $e^{-i\pi\ell\,\mathrm{sign}(\mathrm{Im}\,z')}$. But this allows us a favourable estimate of the integrand on the path C_2 of

Figure 2.2

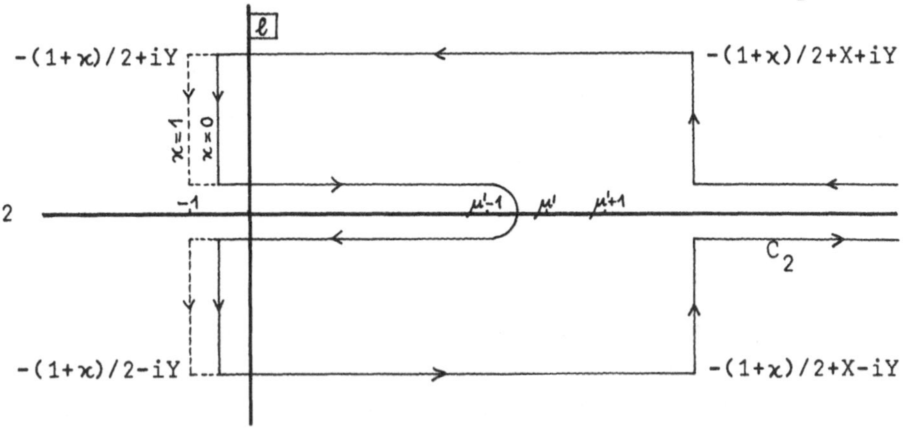

Figure 2.2 into which we may deform C_1 because the integrand has poles at most on the real axis. For if we use the asymptotic expansions (2.6.26) for large $|\ell|$, the integrand essentially is represented as a sum of two exponentials the absolute values of which are given by

(2.6.30)
$$e^{-(\xi'-\xi)X}e^{-(\beta-\beta'+\pi\,\mathrm{sign}\,\beta')Y-\pi|Y|},\quad e^{-(\xi'+\xi)X}e^{-(-\beta-\beta'+\pi\,\mathrm{sign}\,\beta')Y-\pi|Y|},$$
$$z = -\sinh^2\zeta/2,\quad z' = -\sinh^2\zeta'/2,\quad \ell+(1+\varkappa)/2 = X+iY,\quad \zeta,\zeta'\in S_{-\pi,\pi}.$$

For $X \longrightarrow +\infty$ the integrand therefore decreases exponentially for $\xi'>\xi$, i.e. $z\in E(z')$. If we want the same for $Y \longrightarrow \pm\infty$ we must require

(2.6.31)
$$|\beta| < |\beta'|\,.$$

Translated to the z-plane this means that among the hyperbolas with focal points 0 and 1 there must exist one separating z and z' so that z lies on the side of the negative real axis and z' on the other side. Formulated otherwise, z must be contained in $H(z')$, the open part of the plane containing the negative real axis and bounded by the hyperbola through z' with focal points 0 and 1. For $z\in E(z')\cap H(z')$ we therefore may deform the integration path C_2 to the path $-C_3 - C_4$ of Figure 2.3 so that instead of (2.6.28) we get

(2.6.32)
$$\frac{i}{2}\int_{C_3+C_4}\frac{d\ell}{\sin\pi\ell}(2\ell+\varkappa+1)\,u^{\varkappa,\ell}_{\mu'\mu}(z)\,e^{-i\pi\ell\,\mathrm{sign}(\mathrm{Im}\,z')}\,v^{\varkappa,\ell}_{\mu'\mu}(z') =$$
$$= (-)^{\mu'-\mu}\,\frac{1}{2}\,\frac{\omega^{\varkappa}_{\mu'\mu}(z)}{\omega^{\varkappa}_{\mu'\mu}(z')}\,\frac{1}{z-z'}\,,\qquad z\in H(z').$$

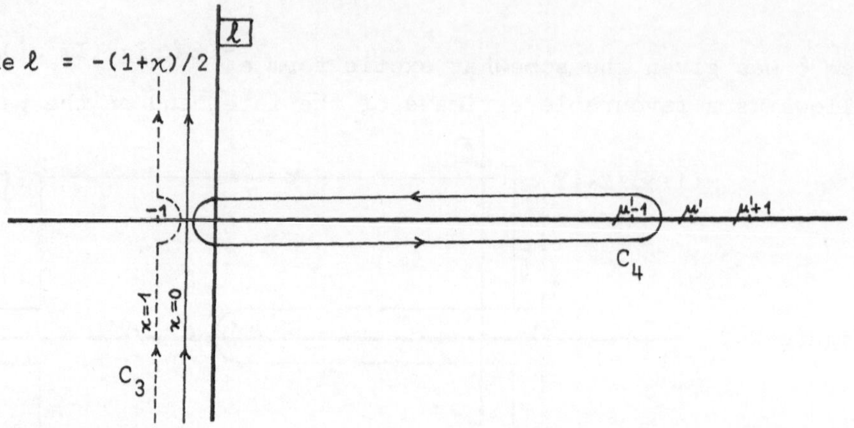

Firgure 2.3

In virtue of its derivation this formula holds for $z \in H(z') \cap E(z')$. But since the integrals exist for all $z \in H(z')$ and represent holomorphic functions there, we can extend the coincidence with the function standing on the right hand side to all $H(z')$. We call (2.6.32) generalized Mehler formula. For with the aid of the relation

$$e^{-i\pi\ell\,\mathrm{sign}(\mathrm{Im}\,z)}v^{\kappa,\ell}_{\mu'\mu}(z) + e^{i\pi\ell\,\mathrm{sign}(\mathrm{Im}\,z)}v^{\kappa,-\ell-\kappa-1}_{\mu'\mu}(z) =$$

(2.6.33)

$$= -(-)^{\mu'-\mu}\pi\,\mathrm{ctg}\,\pi\ell\; e^{i\pi(\mu+\frac{\kappa}{2})\mathrm{sign}(\mathrm{Im}\,\ell)\; -\; i\pi\kappa\,\mathrm{sign}(\mathrm{Im}\,z)/2}\, u^{\kappa,\ell}_{\mu',-\mu-\kappa}(1-z),$$

which follows from (2.6.18) and (2.6.4), the relation (2.6.32) for $\mu' = \mu = \kappa = 0$ is reduced to the well known Mehler formula (cf. [23], p. 174)

$$(2.6.34)\quad -\pi\int_{-\infty}^{+\infty} dp\, \frac{p\,\mathrm{tgh}\pi p}{\cosh\pi p}\, P_{-1/2+ip}(1-2z)\, P_{-1/2+ip}(-(1-2z')) = \frac{1}{z-z'}$$

for the conical functions $P_{-1/2+ip}$.

We use the generalized Heine and Mehler formulas for the derivation of expansion theorems for certain holomorphic functions. Let f be a function of the form

$$f(z) = \omega'^{\kappa}_{\mu'\mu}(z)\; F(z)\;,$$

(2.6.35)

$$\omega'^{\kappa}_{\mu'\mu}(z) := e^{i\pi\,\mathrm{sign}(\mathrm{Im}\,z)(\mu'-\mu)/2}\,\omega^{\kappa}_{\mu'\mu}(z) = z^{(\mu'-\mu)/2}(1-z)^{(\mu'+\mu+\kappa)/2}$$

where F is holomorphic in a domain which contains the interior $E(x_o)$ of an ellipse through $x_o < 0$ with focal points 0 and 1 together with its boundary $\partial E(x_o)$. With the aid of Cauchys integral theorem and observing (2.6.7) we get from (2.6.28)

$$\begin{aligned} f(z) &= \sum_{\ell=\mu'}^{\infty}(2\ell+\varkappa+1)\, \tilde{u}'^{\varkappa,\ell}_{\mu'\mu}(z)\, \hat{f}(\ell)\ , \\ \hat{f}(\ell) &:= \frac{-1}{i\pi}\oint_{\partial E(x_o)} dz\, f(z)\, e^{-i\pi(\mathrm{sign}(\mathrm{Im}\ell)+\mathrm{sign}(\mathrm{Im}\, z))\frac{\mu'-\mu}{2}}\, v^{\varkappa,\ell}_{\mu'\mu}(z)\ . \end{aligned} \tag{2.6.36}$$

If we now let the point z move to a point $x \in [0,1]$ we can contract the integration path $\partial E(x_o)$ around the cut between 0 and 1 of the integrand. Then with the aid of the relation

$$\begin{aligned} & e^{-i\pi\,\mathrm{sign}(\mathrm{Im}\ell)(\mu'-\mu)/2}\Big[e^{i\pi(\mu'-\mu)/2}\, v^{\varkappa,\ell}_{\mu'\mu}(x-io) - \\ & - e^{-i\pi(\mu'-\mu)/2}\, v^{\varkappa,\ell}_{\mu'\mu}(x+io)\Big] = -i\pi\, \tilde{u}'^{\varkappa,\ell}_{\mu'\mu}(x),\ 0 \leq x \leq 1, \end{aligned} \tag{2.6.37}$$

following from (2.6.18) and (2.6.7), instead of (2.6.36) we get

$$\begin{aligned} f(x) &= \sum_{\ell=M}^{\infty}(2\ell+\varkappa+1)\, \tilde{u}'^{\varkappa,\ell}_{\mu'\mu}(x)\, \hat{f}(\ell)\ , \quad 0 \leq x \leq 1\ , \\ \hat{f}(\ell) &:= \int_0^1 dx\, f(x)\, \tilde{u}'^{\varkappa,\ell}_{\mu'\mu}(x)\ , \quad M := \max(\mu',\mu,-\mu'-\varkappa,-\mu-\varkappa)\ , \\ \int_0^1 dx\, |f(x)|^2 &= \sum_{\ell=M}^{\infty}(2\ell+\varkappa+1)\,|\hat{f}(\ell)|^2\ . \end{aligned} \tag{2.6.38}$$

To write this formula in a form valid for all values of μ' and μ we have used here the symmetry relations (2.6.4).

Let g be a function of the form

$$g(z) = \omega^{\varkappa}_{\mu'\mu}(z)\, G(z) \tag{2.6.39}$$

where G is holomorphic in a domain which contains the hyperbola $\partial H(x_o)$ through $x_o \in (0,1)$ with focal points 0 and 1 together with the open part $H(x_o)$ of the complex plane which is separated by $\partial H(x_o)$ and contains the negative real axis. Further let G vanish faster than any power of z for Re z $\longrightarrow -\infty$. Then from (2.6.32) we get with the aid of Cauchys integral theorem

$$\begin{aligned} g(z) &= -\frac{i}{2}\int_{C_3+C_4} d\ell\ (2\ell+\varkappa+1)\ \mathrm{ctg}\pi\ell\ \hat{u}^{\varkappa,\ell}_{\mu'\mu}(z)\, \hat{g}'(\ell)\ , \\ \hat{g}'(\ell) &:= \frac{(-)^{\mu'-\mu}}{i\pi\cos\pi\ell}\int_{\partial H(x_o)} dz\ g(z)\, e^{-i\pi\ell\,\mathrm{sign}(\mathrm{Im}\, z)}\, v^{\varkappa,\ell}_{\mu'\mu}(z)\ . \end{aligned} \tag{2.6.40}$$

The commuting of integrations is allowed because of the presupposed asymptotic behaviour of G. Observe that the integral over $\partial H(x_o)$ does not go over a closed path because of the cut from $v^{\varkappa,\ell}_{\mu'\mu}$ in the interval (0,1). If we let z move to a point $x \in (-\infty,0]$ we can contract the integration path $\partial H(x_0)$ around the negative real axis. Since $v^{\varkappa,\ell}_{\mu'\mu}$ is continuous there, we get from (2.6.40)

$$(2.6.41)\quad \begin{aligned} g(x) &= -\frac{i}{2}\int_{C_3+C_4} d\ell\,(2\ell+\varkappa+1)\,\mathrm{ctg}\pi\ell\;\hat{u}^{\varkappa,\ell}_{\mu'\mu}(x)\;\hat{g}'(\ell),\quad -\infty<x\leq 0\,,\\ \hat{g}'(\ell) &:= \frac{2}{\pi}(-)^{\mu'-\mu}\,\mathrm{tg}\pi\ell\int_{-\infty}^{0} dx\; g(x)\; v^{\varkappa,\ell}_{\mu'\mu}(x)\,. \end{aligned}$$

Because of the zero of the factor $2\ell+\varkappa+1$ we may choose as the path C_3 in Figure 2.3 a straight line through the point $\ell = -(1+\varkappa)/2$ also in the case $\varkappa = 1$. Therefore in any case the substitution $\ell \longrightarrow -\ell-\varkappa-1$ takes C_3 into $-C_3$. Because of the symmetry (2.6.4) of $u^{\varkappa,\ell}_{\mu'\mu}$ under this substitution only the symmetrical part

$$(2.6.42)\qquad \hat{g}(\ell) := \frac{\hat{g}'(\ell) - \hat{g}'(-\ell-\varkappa-1)}{2}$$

contributes to the C_3-part of the integral in (2.6.41). The relation

$$(2.6.43)\qquad v^{\varkappa,\ell}_{\mu'\mu}(z) - v^{\varkappa,-\ell-\varkappa-1}_{\mu'\mu}(z) = (-)^{\mu'-\mu}\pi\,\mathrm{ctg}\pi\ell\; u^{\varkappa,\ell}_{\mu'\mu}(z)$$

following from (2.6.18) yields

$$(2.6.44)\qquad \hat{g}(\ell) = \int_{-\infty}^{0} dx\; g(x)\;\hat{u}^{\varkappa,\ell}_{\mu'\mu}(x)\,.$$

At the integer points $\ell = 0,1,\ \dots\ ,\mu'-1$ which are enclosed by the path C_4 the function $u^{\varkappa,\ell}_{\mu'\mu}(z)\; v^{\varkappa,\ell}_{\mu'\mu}(z')$ according to (2.6.1) and (2.6.15) has poles at the points $\ell = 0,1,\ \dots\ ,\mu-1$ for $\mu \geq 1$ and is regular in all other possible cases. From (2.6.18) at once follows

$$(2.6.45)\qquad \operatorname*{Res}_{\ell\in\{0,1,\dots,\mu-1\}} v^{\varkappa,\ell}_{\mu'\mu}(z) = \frac{(-)^{\mu'-\mu}}{2}\, u^{\varkappa,\ell}_{\mu'\mu}(z)\,,\quad \mu\geq 1\,,$$

because $u^{\varkappa,\ell}_{\mu',-\mu-\varkappa}(1-z)$ has zeros at these points. For $\ell \in \{0,1,\dots,\mu'-1\}$ we therefore get

$$(2.6.46)\quad \hat{g}'(\ell) = \hat{g}(\ell) = \begin{cases} \displaystyle\int_{-\infty}^{0} dx\; g(x)\;\hat{u}^{\varkappa,\ell}_{\mu'\mu}(x) & \text{for } \ell = 0,1,\dots,\mu-1;\ \mu\geq 1,\\[2ex] 0 & \text{otherwise}\,. \end{cases}$$

Summarizing the last results instead of (2.6.41) we can write

(2.6.47)
$$g(x) = \frac{1}{2\pi i}\int_{-(1+\varkappa)/2-i\infty}^{-(1+\varkappa)/2+i\infty} d\ell\ (2\ell+\varkappa+1)\pi\,\mathrm{ctg}\pi\ell\ \hat{u}^{\varkappa,\ell}_{\mu'\mu}(x)\ \hat{g}(\ell) + \sum_{\ell=0}^{N}(2\ell+\varkappa+1)\ \hat{u}^{\varkappa,\ell}_{\mu'\mu}(x)\ \hat{g}(\ell)\ , \qquad -\infty < x \leq 0\ ,$$
$$\hat{g}(\ell) := \int_{-\infty}^{0} dx\ g(x)\ \hat{u}^{\varkappa,\ell}_{\mu'\mu}(x)\ ,$$
$$\int_{-\infty}^{0} dx\,|g(x)|^2 = \frac{1}{2\pi i}\int_{-(1+\varkappa)/2-i\infty}^{-(1+\varkappa)/2+i\infty} d\ell\ (2\ell+\varkappa+1)\pi\,\mathrm{ctg}\pi\ell\,|\hat{g}(\ell)|^2 + \sum_{\ell=0}^{N}(2\ell+\varkappa+1)\,|\hat{g}(\ell)|^2\ ,$$
$$N := M - |\mu'-\mu| - 1\ ,\quad M := \max(\mu',\mu,-\mu'-\varkappa,-\mu-\varkappa)\ .$$

Here again the symmetry relations (2.6.4) are used to write the result in a form valid for all values of μ',μ. The sum is to be interpreted as 0 for $N<0$.

To the expansion formulas (2.6.38) and (2.6.47) we may derive dual relations which essentially contain the orthogonality relations for the functions $u^{\varkappa,\ell}_{\mu'\mu}$. We first use the differential equations

(2.6.48)
$$\left[\Delta^{\varkappa,\ell}_{\mu'\mu} + (\ell+\varkappa/2)(\ell+\varkappa/2+1)\right] u^{\varkappa,\ell}_{\mu'\mu}(z) = 0\ ,$$
$$\left[\Delta^{\varkappa,\ell}_{\mu'\mu} + (\ell+\varkappa/2)(\ell+\varkappa/2+1)\right] v^{\varkappa,\ell}_{\mu'\mu}(z) = 0\ ,$$
$$\Delta^{\varkappa,\ell}_{\mu'\mu} := z(1-z)\frac{d^2}{dz^2} + (1-2z)\frac{d}{dz} - \frac{(\mu'-\mu)^2 + (2\mu'+\varkappa)(2\mu+\varkappa)z}{4z(1-z)}\ ,$$

which are equivalent to the corresponding hypergeometric differential equations, to get dual relations to the generalized Heine and Mehler formulas. The step operations

(2.6.49)
$$\frac{\Lambda^{\varkappa,\ell,+}_{\mu'-1,\mu}\ w^{\varkappa,\ell}_{\mu'-1,\mu}(z)}{(\ell-\mu'+1)(\ell+\mu'+\varkappa)N^{\varkappa,\ell}_{\mu'-1,\mu}} = \frac{w^{\varkappa,\ell}_{\mu'\mu}(z)}{N^{\varkappa,\ell}_{\mu'\mu}} = \frac{\Lambda^{\varkappa,\ell,-}_{\mu'+1,\mu}\ w^{\varkappa,\ell}_{\mu'+1,\mu}(z)}{N^{\varkappa,\ell}_{\mu'+1,\mu}}\ ,$$
$$\Lambda^{\varkappa,\ell,\pm}_{\mu'\mu} := \pm\sqrt{-z(1-z)}\ \frac{d}{dz} + \frac{(\mu'-\mu)(1-z)-(\mu'+\mu+\varkappa)z}{2\sqrt{-z(1-z)}}\ ,\quad w^{\varkappa,\ell}_{\mu'\mu} = \begin{cases} u^{\varkappa,\ell}_{\mu'\mu} & \text{or} \\ v^{\varkappa,\ell}_{\mu'\mu}\ , \end{cases}$$

correspond to well known differentiation formulas for hypergeometric functions. With these the validity of the relation

$$u^{\varkappa,\ell}_{\mu'\mu}(z)\, v^{\varkappa,\ell'}_{\mu'\mu}(z)\, e^{-i\pi\ell'\,\mathrm{sign}(\mathrm{Im}\, z)} = \frac{-\,e^{-i\pi\ell'\,\mathrm{sign}(\mathrm{Im}\, z)}}{(\ell+\ell'+\varkappa+1)(\ell-\ell')} \times$$

(2.6.50)

$$\times \frac{d}{dz}\left\{\sqrt{-z(1-z)}\left[\frac{N^{\varkappa,\ell}_{\mu'\mu}}{N^{\varkappa,\ell}_{\mu'-1,\mu}} v^{\varkappa,\ell'}_{\mu'\mu}(z)\, u^{\varkappa,\ell}_{\mu'-1,\mu}(z) + \frac{N^{\varkappa,\ell'}_{\mu'\mu}}{N^{\varkappa,\ell'}_{\mu'-1,\mu}} v^{\varkappa,\ell'}_{\mu'-1,\mu}(z)\, u^{\varkappa,\ell}_{\mu'\mu}(z)\right]\right\}$$

can be traced back to the validity of the differential equation (2.6.48). Group theoretically the $\Lambda^{\varkappa,\ell,\pm}_{\mu'\mu}$ and $\Delta^{\varkappa,\ell}_{\mu'\mu} = \Lambda^{\varkappa,\ell,\pm}_{\mu'\mp1,\mu}\Lambda^{\varkappa,\ell,\mp}_{\mu'\mu} - (\mu'+\varkappa/2)(\mu'+\varkappa/2\mp1)$ mean the differential operators that are associated with the well known step operators of the Lie algebra and the Casimir operator, respectively, of the groups SU(2) or SU(1,1). The function on the left hand side in (2.6.50) is analytical and single-valued in the z-plane cut along the real axis. Now we integrate (2.6.50) over the open semiarcs lying in the upper and the lower half-plane of the positively oriented ellipse $\partial E(x_o)$, $x_o<0$ (cf. before (2.6.36)). With the relation

$$e^{i\pi((\mu'-\mu)/2+\ell)} v^{\varkappa,\ell}_{\mu'\mu}(x-io) - e^{-i\pi((\mu'-\mu)/2+\ell)} v^{\varkappa,\ell}_{\mu'\mu}(x+io) =$$

(2.6.51)

$$= i\pi(-)^{\mu+\varkappa+1}\, e^{-i\pi\,\mathrm{sign}(\mathrm{Im}\ell)(\mu'-\mu)/2}\, \tilde{u}'^{\varkappa,\ell}_{\mu',-\mu-\varkappa}(1-x),\quad 0<x<1\ ,$$

following from (2.6.18) the left hand side with (2.6.7) in the limit $x_o \longrightarrow -0$ yields the integral

$$(2.6.52)\quad i\pi(-)^{\mu+\varkappa+1} e^{i\pi(\mathrm{sign}(\mathrm{Im}\ell)-\mathrm{sign}(\mathrm{Im}\ell'))\frac{\mu'-\mu}{2}} \int_0^1 dx\, \tilde{u}'^{\varkappa,\ell}_{\mu'\mu}(x)\, \tilde{u}'^{\varkappa,\ell'}_{\mu',-\mu-\varkappa}(1-x).$$

The right hand side is calculated from the values of the curled bracket at the end points of the integration path x_o-io, $1-x_o$+io, $1-x_o$-io, x_o+io. In the limit $x_o \longrightarrow -0$ we get

$$(2.6.53)\quad \frac{-i(-)^{\mu'-\mu}}{(\ell+\ell'+\varkappa+1)(\ell-\ell')} \frac{N^{\varkappa,\ell}_{\mu'\mu}}{N^{\varkappa,\ell'}_{\mu'\mu}} \frac{\Gamma(1+\ell'-\mu)}{\Gamma(1+\ell'+\mu+\varkappa)}\left[\frac{\Gamma(1+\ell+\mu+\varkappa)}{\Gamma(1+\ell-\mu)}\sin\pi\ell - \frac{\Gamma(1+\ell'+\mu+\varkappa)}{\Gamma(1+\ell'-\mu)}\sin\pi\ell'\right].$$

The last two formulas therefore yield the relation

$$(\ell'+\ell+\varkappa+1)\int_0^1 dx\, \tilde{u}'^{\varkappa,\ell}_{\mu'\mu}(x)\, (-)^{\mu'+\varkappa}\, \tilde{u}'^{\varkappa,\ell'}_{\mu',-\mu-\varkappa}(1-x) =$$

$$(2.6.54)\quad = e^{-i\pi(\mathrm{sign}(\mathrm{Im}\ell)-\mathrm{sign}(\mathrm{Im}\ell'))(\mu'-\mu)/2} \frac{N^{\varkappa,\ell}_{\mu'\mu}}{N^{\varkappa,\ell'}_{\mu'\mu}} \frac{1}{\pi(\ell-\ell')} \times$$

$$\times \frac{\Gamma(1+\ell'-\mu)}{\Gamma(1+\ell'+\mu+\varkappa)}\left[\frac{\Gamma(1+\ell+\mu+\varkappa)}{\Gamma(1+\ell-\mu)}\sin\pi\ell - \frac{\Gamma(1+\ell'+\mu+\varkappa)}{\Gamma(1+\ell'-\mu)}\sin\pi\ell'\right].$$

This can be interpreted as dual to the generalized Heine formula

(2.6.28). From (2.6.18) follows for integer $\ell \geq \mu', -\mu'-\varkappa, \mu, -\mu-\varkappa$

(2.6.55) $$(-)^{\mu'+\varkappa}\, \tilde{u}'^{\varkappa,\ell}_{\mu',-\mu-\varkappa}(1-x) = (-)^{\ell}\, \tilde{u}'^{\varkappa,\ell}_{\mu'\mu}(x) .$$

With this and (2.6.2) one derives at once the orthogonality relations

(2.6.56) $$\int_0^1 dx\, \tilde{u}'^{\varkappa,\ell}_{\mu'\mu}(x)\, \tilde{u}'^{\varkappa,\ell'}_{\mu'\mu}(x) = \frac{\delta_{\ell\ell'}}{2\ell+\varkappa+1} ,$$
$$\ell,\ell' \geq M, \text{ integer} , \quad M: = \max(\mu',\mu,-\mu'-\varkappa,-\mu-\varkappa) .$$

For any complex valued function $\hat{f}$ on the point set $\{M,M+1,M+2, \ldots\}$ with compact support therefore holds

(2.6.57) $$\hat{f}(\ell) = \int_0^1 dx\, \tilde{u}'^{\varkappa,\ell}_{\mu'\mu}(x)\, f(x) , \qquad \ell \in \{M,M+1,M+2, \ldots\} ,$$
$$f(x): = \sum_{\ell=M}^{\infty} (2\ell+\varkappa+1)\, \tilde{u}'^{\varkappa,\ell}_{\mu'\mu}(x)\, \hat{f}(\ell), \; x \in [0,1] ,$$
$$\int_0^1 dx\, |f(x)|^2 = \sum_{\ell=M}^{\infty} (2\ell+\varkappa+1)\, |\hat{f}(\ell)|^2 .$$

If we integrate (2.6.50) over the semiarcs lying in the lower and the upper half-plane, respectively, of the positively oriented hyperbola $\partial H(x_o)$, $x_o \in (0,1)$ (cf. before (2.6.40)), the integral converges only for $\mathrm{Re}(\ell'+(1+\varkappa)/2) > |\mathrm{Re}(\ell+(1+\varkappa)/2)|$. In this case the left hand side yields in the limit $x_o \longrightarrow +0$

(2.6.58) $$2i \sin\pi\ell' \int_{-\infty}^{0} dx\, u^{\varkappa,\ell}_{\mu'\mu}(x)\, v^{\varkappa,\ell'}_{\mu'\mu}(x) .$$

The right hand side becomes the difference of the values of the curled bracket at the integration end points, yielding in the limit $x_o \longrightarrow +0$

(2.6.59) $$i \sin\pi\ell' (-)^{\mu'-\mu}\, N^{\varkappa,\ell}_{\mu'\mu} / N^{\varkappa,\ell'}_{\mu'\mu}\, \frac{1}{(\ell+\ell'+\varkappa+1)(\ell'-\ell)} .$$

The final result is the formula

(2.6.60) $$\int_{-\infty}^{0} dx\, u^{\varkappa,\ell}_{\mu'\mu}(x)\, v^{\varkappa,\ell'}_{\mu'\mu}(x) = \frac{1}{2}\, (-)^{\mu'-\mu}\, N^{\varkappa,\ell}_{\mu'\mu} / N^{\varkappa,\ell'}_{\mu'\mu}\, \frac{1}{(\ell'+\ell+\varkappa+1)(\ell'-\ell)} ,$$
$$\mathrm{Re}(\ell'+(1+\varkappa)/2) > |\mathrm{Re}(\ell+(1+\varkappa)/2)| ,$$

which can be interpreted as dual to the generalized Mehler formula

(2.6.32). Now let be $\mu \geq 1$ and $\ell \in \{0,1, \ldots ,\mu-1\}$. If ℓ' moves to an integer $\geq \ell$ out of the set $\{0,1, \ldots ,\mu-1\}$, from (2.6.60) with (2.6.45) at once follow the orthogonality relations

$$\int_{-\infty}^{0} dx\, \hat{u}^{\varkappa,\ell}_{\mu'\mu}(x)\, \hat{u}^{\varkappa,\ell'}_{\mu'\mu}(x) = \frac{\delta_{\ell\ell'}}{2\ell+\varkappa+1}, \qquad \ell,\ell' \in \{0,1, \ldots ,N\},$$

(2.6.61)

$$N: = M-|\mu'-\mu|-1 \;, \; N \geq 0 \;, \; M: = \max(\mu',\mu,-\mu'-\varkappa,-\mu-\varkappa) \;.$$

Here the restriction $\ell' \geq \ell$ may be omitted because of the symmetric form of the relation; the validity for all μ',μ with $N \geq 0$ follows from the symmetry relations (2.6.4). Let $\hat{g}$ be of the form $\hat{g}(\ell) = N^{\varkappa,\ell}_{\mu'\mu}\, \hat{G}(\ell)$ with $\hat{G}$ an entire function, vanishing for $\mathrm{Im}\,\ell \longrightarrow \pm\infty$ faster than any power and symmetrical under the substitution $\ell \longrightarrow -\ell-\varkappa-1$. Choose $\ell \in$ $\in -(1+\varkappa)/2 + i\mathbb{R} = C_3$ and integrate the relation (2.6.60), multiplied by $(2\ell'+\varkappa+1)\hat{g}(\ell')$, along $C_3+\varepsilon$ upwards and along $C_3-\varepsilon$ downwards $(0<\varepsilon<\frac{1}{2})$ where in the second integral because of the restriction for $\mathrm{Re}\,\ell'$ $v^{\varkappa,\ell'}_{\mu'\mu}$ is substituted by $v^{\varkappa,-\ell'-\varkappa-1}_{\mu'\mu}$. On the right hand side the sum of the residues gives $i\pi(-)^{\mu'-\mu}[\hat{g}(\ell) + \hat{g}(-\ell-\varkappa-1)] = 2\pi i\, (-)^{\mu'-\mu}\, \hat{g}(\ell)$ while the left hand side yields the integral

$$\int_{-\infty}^{0} dx\, u^{\varkappa,\ell}_{\mu'\mu}(x) \int_{-(1+\varkappa)/2-i\infty}^{-(1+\varkappa)/2+i\infty} d\ell' \Big[(2\ell'+\varkappa+1+2\varepsilon)\, v^{\varkappa,\ell'+\varepsilon}_{\mu'\mu}(x)\, \hat{g}(\ell'+\varepsilon) -$$

(2.6.62)

$$- (2\ell'+\varkappa+1-2\varepsilon)\, v^{\varkappa,-\ell'-\varkappa-1+\varepsilon}_{\mu'\mu}(x)\, \hat{g}(\ell'-\varepsilon)\Big].$$

The integrations may be commuted because of the asymptotic behaviour of the functions. Since the right hand side does not depend on ε because of continuity in the limit $\varepsilon \longrightarrow 0$ one gets with (2.6.43) the generalized orthogonality relations

$$\hat{g}(\ell) = \int_{-\infty}^{0} dx\, \hat{u}^{\varkappa,\ell}_{\mu'\mu}(x)\, g(x) \;, \qquad \ell \in C_3 = -(1+\varkappa)/2 + i\mathbb{R} \;,$$

(2.6.63) $$g(x): = \frac{1}{2\pi i} \int_{C_3} d\ell\; (2\ell+\varkappa+1)\pi\, \mathrm{ctg}\,\pi\ell\; \hat{u}^{\varkappa,\ell}_{\mu'\mu}(x)\, \hat{g}(\ell) \;, \quad x \leq 0 \;,$$

$$\int_{-\infty}^{0} dx\, |g(x)|^2 = \frac{1}{2\pi i} \int_{C_3} d\ell\; (2\ell+\varkappa+1)\pi\, \mathrm{ctg}\,\pi\ell\, |\hat{g}(\ell)|^2 \;.$$

Finally form (2.6.60) with (2.6.45) for $N \geq 0$ also follows

(2.6.64) $$\int_{-\infty}^{0} dx\, \hat{u}^{\varkappa,\ell}_{\mu'\mu}(x)\, \hat{u}^{\varkappa,\ell'}_{\mu'\mu}(x) = 0 \;, \qquad \ell \in C_3 \;, \quad \ell' \in \{0,1, \ldots ,N\} \;.$$

Obviously the formulas (2.6.61) to (2.6.64) may be combined to

$$\hat{g}(\ell) = \int_{-\infty}^{0} dx\, \hat{u}^{\varkappa,\ell}_{\mu'\mu}(x)\, g(x)\ , \quad \ell \in C_3 \cup \{0,1,\ \dots\ ,N\}\ ,$$

(2.6.65)

$$g(x) := \frac{1}{2\pi i} \int_{C_3} d\ell\ (2\ell+\varkappa+1)\pi\, \mathrm{ctg}\pi\ell\ \hat{u}^{\varkappa,\ell}_{\mu'\mu}(x)\ \hat{g}(\ell) + \sum_{\ell=0}^{N} (2\ell+\varkappa+1)\ \hat{u}^{\varkappa,\ell}_{\mu'\mu}(x)\ \hat{g}(\ell)\ ,$$

$$\int_{-\infty}^{0} dx\, |g(x)|^2 = \frac{1}{2\pi i} \int_{C_3} d\ell\ (2\ell+\varkappa+1)\pi\, \mathrm{ctg}\pi\ell\, |\hat{g}(\ell)|^2 + \sum_{\ell=0}^{N} (2\ell+\varkappa+1)\, |\hat{g}(\ell)|^2 .$$

It can be shown that the expansions (2.6.38) and (2.6.47) as well as the dual expansions (2.6.57) and (2.6.65) can be completed to unitary maps between the Hilbert spaces $\mathcal{L}^2(0,1)$ and $\mathcal{L}^2(-\infty,0)$, respectively, and the Hilbert spaces $\ell^{2,\varkappa}_{\mu'\mu}$ with scalar product

(2.6.66)
$$\langle \hat{f}' | \hat{f} \rangle^{\wedge\varkappa}_{\mu'\mu} := \sum_{\ell=M}^{\infty} (2\ell+\varkappa+1)\ \hat{f}'(\ell)^*\, \hat{f}(\ell)$$

and $\mathcal{H}^{\varkappa}_{\mu'\mu}$ with scalar product

(2.6.67)

$$\langle \hat{g}' | \hat{g} \rangle^{\wedge\varkappa}_{\mu'\mu} := \frac{1}{2\pi i} \int_{C_3} d\ell\ (2\ell+\varkappa+1)\pi\, \mathrm{ctg}\pi\ell\ \hat{g}'(\ell)^*\, \hat{g}(\ell) + \sum_{\ell=0}^{N} (2\ell+\varkappa+1)\ \hat{g}'(\ell)^*\, \hat{g}(\ell)\ ,$$

respectively. In the first case the assertion follows from the fact that the systems $\{f_\ell :\ \ell = M, M+1,\ \dots\ ;\ f_\ell(x) = \sqrt{2\ell+\varkappa+1}\ \tilde{u}'^{\varkappa,\ell}_{\mu'\mu}(x)\}$ and $\{\hat{f}_\ell :\ \ell = M, M+1,\ \dots\ ;\ \hat{f}_\ell(\ell') = \delta_{\ell\ell'}/\sqrt{2\ell+\varkappa+1}\}$ fulfil the assumptions under which the expansions (2.6.38) and (2.6.57), respectively, were derived. Here $\hat{f}_\ell$ is the image of f_ℓ and vice versa. As shown with the approximation theorem of Weierstraß the system $\{f_\ell\}$ is a complete orthonotmal basis in $\mathcal{L}^2(0,1)$ while $\{\hat{f}_\ell\}$ obviously represents such a basis in $\ell^{2,\varkappa}_{\mu'\mu}$. In the other case a proof with only simple analytical means is not known to us so that we omit it here. In the following the pairs of formulas (2.6.38), (2.6.57) and (2.6.47), (2.6.65) are considered as unitary transformations and completed to the Fourier analysis in the Hilbert spaces $\mathcal{L}^2(SU(2))$ and $\mathcal{L}^2(SU(1,1))$ of square-integrable functions on the groups SU(2) and SU(1,1), respectively.

The invariant Haar measure on both groups in the parametrization defined by (2.6.11) and (2.6.13) is

(2.6.68)
$$\int dA := \int \frac{d\alpha}{2\pi}\, dz\, \frac{d\gamma}{4\pi}$$

Here $z \in [0,1]$ for $A \in SU(2)$ and $z \in (-\infty,0]$ for $A \in SU(1,1)$. The normalization is choosen so that we have $\int_{SU(2)} dA = 1$ for the compact group SU(2). Each element $f \in \mathcal{L}^2(G)$, $G \in \{SU(2), SU(1,1)\}$ can be written as a function of the parameters α, z, γ. If f is continued to a periodical function in α and γ by the condition

$$f(A) = f(\alpha,z,\gamma) = f(\alpha,z,\gamma+4\pi) = f(\alpha+2\pi,z,\gamma+2\pi) , \tag{2.6.69}$$

we get by expansion into a Fourier series with respect to γ

$$\begin{aligned} f(\alpha,z,\gamma) &= \sum_{\varkappa=0,1} \sum_{\mu=-\infty}^{+\infty} f_{\mu+\varkappa/2}(\alpha,z)\, e^{-i(\mu+\varkappa/2)\gamma} , \\ f_{\mu+\varkappa/2}(\alpha,z) &= \int_0^{4\pi} \frac{d\gamma}{4\pi}\, e^{i(\mu+\varkappa/2)\gamma} f(\alpha,z,\gamma) , \end{aligned} \tag{2.6.70}$$

where because of the periodicity condition (2.6.69) we have

$$f_{\mu+\varkappa/2}(\alpha+2\pi,z) = (-)^{\varkappa} f_{\mu+\varkappa/2}(\alpha,z) . \tag{2.6.71}$$

For $f_{\mu+\varkappa/2}$ we therefore get the Fourier expansion with respect to

$$\begin{aligned} f_{\mu+\varkappa/2}(\alpha,z) &= \sum_{\mu'=-\infty}^{+\infty} f^{\varkappa}_{\mu'\mu}(z)\, e^{-i(\mu'+\varkappa/2)\alpha} , \\ f^{\varkappa}_{\mu'\mu}(z) &= \int_0^{2\pi} \frac{d\alpha}{2\pi}\, e^{i(\mu'+\varkappa/2)\alpha} f_{\mu+\varkappa/2}(\alpha,z) . \end{aligned} \tag{2.6.72}$$

If for $f^{\varkappa}_{\mu'\mu}(z)$ we finally take the expansions (2.6.38) or (2.6.47), respectively, we get

$$f(\alpha,x,\gamma) = \sum_{\varkappa=0,1} \sum_{\mu',\mu=-\infty}^{+\infty} \sum_{\ell=M}^{\infty} (2\ell+\varkappa+1)\, \hat{f}^{\varkappa,\ell}_{\mu'\mu} \times e^{-i(\mu'+\varkappa/2)\alpha}\, \tilde{u}^{\prime\varkappa,\ell}_{\mu'\mu}(x)\, e^{-i(\mu+\varkappa/2)\gamma} ,$$

$$\hat{f}^{\varkappa,\ell}_{\mu'\mu} = \int_0^{2\pi} \frac{d\alpha}{2\pi} \int_0^1 dx \int_0^{4\pi} \frac{d\gamma}{4\pi}\, e^{i(\mu'+\varkappa/2)\alpha}\, \tilde{u}^{\prime\varkappa,\ell}_{\mu'\mu}(x)\, e^{i(\mu+\varkappa/2)\gamma} f(\alpha,x,\gamma) , \tag{2.6.73}$$

$$0 \leq x \leq 1, \quad \ell \in \{M, M+1, \dots\} , \quad M := \max(\mu',\mu,-\mu'-\varkappa,-\mu-\varkappa) ;$$

$$f(\alpha,x,\gamma) = \sum_{\varkappa=0,1} \sum_{\mu',\mu=-\infty}^{+\infty} \Big\{ \frac{1}{i\pi} \int_{-(1+\varkappa)/2}^{-(1+\varkappa)/2+i\infty} d\ell\, (2\ell+\varkappa+1)\pi \operatorname{ctg}\pi\ell\, \hat{f}^{\varkappa,\ell}_{\mu'\mu} \times e^{-i(\mu'+\varkappa/2)\alpha}\, \hat{u}^{\varkappa,\ell}_{\mu'\mu}(x)\, e^{-i(\mu+\varkappa/2)\gamma} +$$

$$+ \sum_{\ell=0}^{N} (2\ell+\varkappa+1)\, \hat{f}^{\varkappa,\ell}_{\mu'\mu}\, e^{-i(\mu'+\varkappa/2)\alpha}\, \hat{u}^{\varkappa,\ell}_{\mu'\mu}(x)\, e^{-i(\mu+\varkappa/2)\gamma} \Big\},$$

$$(2.6.73)\quad \hat{f}^{\varkappa,\ell}_{\mu'\mu} = \int_0^{2\pi}\frac{d\alpha}{2\pi} \int_{-\infty}^{0} dx \int_0^{4\pi}\frac{d\gamma}{4\pi}\, e^{i(\mu'+\varkappa/2)\alpha}\, \hat{u}^{\varkappa,\ell}_{\mu'\mu}(x)\, e^{i(\mu+\varkappa/2)\gamma}\, f(\alpha,x,\gamma),$$

$$-\infty < x \le 0,\ \ \ell \in \{-(1+\varkappa)/2+i[0,+\infty)\} \cup \{0,1,\ldots,N\},\ \ N:=M-|\mu'-\mu|-1.$$

Here the last sum must be omitted for $N<0$. Notice that the $\tilde{u}^{|\varkappa,\ell}_{\mu'\mu}$ and $\hat{u}^{\varkappa,\ell}_{\mu'\mu}$ are real functions so that (2.6.73), after a reordering of the summations, with (2.2.4) and (2.3.15) may be rewritten

$$f(A) = \sum_{\varkappa=0,1} \sum_{\ell=0}^{\infty} (2\ell+\varkappa+1) \sum_{\mu',\mu=-\ell-\varkappa}^{\ell} \hat{f}^{\varkappa,\ell}_{\mu'\mu}\, U^{\varkappa,\ell}_{SU(2)}(A)^{\dagger}_{\mu\mu'},$$

$$\hat{f}^{\varkappa,\ell}_{\mu'\mu} = \int_{SU(2)} dA\, U^{\varkappa,\ell}_{SU(2)}(A)_{\mu'\mu}\, f(A),$$

$$A \in SU(2),\ \ \ell \in \{0,1,\ldots\};$$

$$f(A) = \sum_{\varkappa=0,1} \Big\{ \frac{1}{i\pi} \int_{-(1+\varkappa)/2}^{-(1+\varkappa)/2+i\infty} d\ell\, (2\ell+\varkappa+1)\pi \operatorname{ctg}\pi\ell \times$$

$$(2.6.74)\quad \times \sum_{\mu',\mu=-\infty}^{+\infty} \hat{f}^{\varkappa,\ell,o}_{\mu'\mu}\, U^{\varkappa,\ell,o}_{SU(1,1)}(A)^{\dagger}_{\mu\mu'} +$$

$$+ \sum_{\ell=0}^{\infty} (2\ell+\varkappa+1) \Big[\sum_{\mu',\mu=\ell+1}^{+\infty} \hat{f}^{\varkappa,\ell,+}_{\mu'\mu}\, U^{\varkappa,\ell,+}_{SU(1,1)}(A)^{\dagger}_{\mu\mu'} +$$

$$+ \sum_{\mu',\mu=-\ell-\varkappa-1}^{-\infty} \hat{f}^{\varkappa,\ell,-}_{\mu'\mu}\, U^{\varkappa,\ell,-}_{SU(1,1)}(A)^{\dagger}_{\mu\mu'} \Big] \Big\},$$

$$\hat{f}^{\varkappa,\ell,\eta}_{\mu'\mu} = \int_{SU(1,1)} dA\, U^{\varkappa,\ell,\eta}_{SU(1,1)}(A)_{\mu'\mu}\, f(A),$$

$$A \in SU(1,1),\ \ \ell \in \{-(1+\varkappa)/2+i[0,+\infty)\} \cup \{0,1,\ldots\}.$$

The orthogonality and completeness relations for the Fourier series (2.6.70) and (2.6.72) and for the expansions (2.6.38) and (2.6.47) can be combined to the Plancherel formulas for SU(2) and SU(1,1)

$$\int_{SU(2)} dA\,|f(A)|^2 = \sum_{\varkappa=0,1} \sum_{\ell=0}^{\infty} (2\ell+\varkappa+1) \sum_{\mu',\mu=-\ell-\varkappa}^{\ell} |\hat{f}^{\varkappa,\ell}_{\mu'\mu}|^2,$$

$$(2.6.75)\quad \int_{SU(1,1)} dA\,|f(A)|^2 = \sum_{\varkappa=0,1} \Big\{ \frac{1}{i\pi} \int_{-(1+\varkappa)/2}^{-(1+\varkappa)/2+i\infty} d\ell\, (2\ell+\varkappa+1)\pi \operatorname{ctg}\pi\ell \times$$

$$\times \sum_{\mu',\mu=-\infty}^{+\infty} |\hat{f}^{\varkappa,\ell,o}_{\mu'\mu}|^2 +$$

$$(2.6.75) \qquad + \sum_{\ell=0}^{\infty}(2\ell+\varkappa+1)\Big[\sum_{\mu',\mu=\ell+1}^{+\infty}|\hat{f}^{\varkappa,\ell,+}_{\mu'\mu}|^2 + \sum_{\mu',\mu=-\ell-\varkappa-1}^{-\infty}|\hat{f}^{\varkappa,\ell,-}_{\mu'\mu}|^2\Big]\Big\}.$$

They must be interpreted as the unitarity relations for the transformations defined by (2.6.74) between the Hilbert spaces $\mathcal{L}^2(G)$ and the Hilbert spaces $\mathcal{L}^2(\hat{G})$ with the scalar products defined by the right hand sides of (2.6.75). In this connection we emphasize that the expansions (2.6.74) do not converge pointwise but only with respect to the norms of the corresponding pairs of Hilbert spaces. There exist, however, dense subspaces in $\mathcal{L}^2(G)$ and $\mathcal{L}^2(\hat{G})$, respectively, on which the convergence can be considered as pointwise. Still another notation of the formulas (2.6.74) and (2.6.75) is

$$f(A) = \sum_{\varkappa=0,1}\sum_{\ell=0}^{\infty}(2\ell+\varkappa+1)\ \mathrm{Sp}(\hat{f}^{\varkappa,\ell}\ U^{\varkappa,\ell}_{SU(2)}(A)^{\dagger}), \quad A\in SU(2),$$

$$\hat{f}^{\varkappa,\ell} = \int_{SU(2)} dA\ U^{\varkappa,\ell}_{SU(2)}(A)\ f(A), \qquad \ell\in\{0,1,\ldots\},$$

$$\int_{SU(2)} dA\,|f(A)|^2 = \sum_{\varkappa=0,1}\sum_{\ell=0}^{\infty}(2\ell+\varkappa+1)\ \mathrm{Sp}(\hat{f}^{\varkappa,\ell\dagger}\hat{f}^{\varkappa,\ell}) \quad ;$$

$$f(A) = \sum_{\varkappa=0,1}\Big\{\frac{1}{i\pi}\int_{-(1+\varkappa)/2}^{-(1+\varkappa)/2+i\infty} d\ell\ \ (2\ell+\varkappa+1)\,\pi\,\mathrm{ctg}\,\pi\ell\ \times$$

$$\times\ \mathrm{Sp}(\hat{f}^{\varkappa,\ell,o}\ U^{\varkappa,\ell,o}_{SU(1,1)}(A)^{\dagger}) +$$

$$(2.6.76) \qquad + \sum_{\ell=0}^{\infty}(2\ell+\varkappa+1)\sum_{\eta=\pm}\mathrm{Sp}(\hat{f}^{\varkappa,\ell,\eta}\ U^{\varkappa,\ell,\eta}_{SU(1,1)}(A)^{\dagger})\Big\}, \quad A\in SU(1,1),$$

$$\hat{f}^{\varkappa,\ell,\eta} = \int_{SU(1,1)} dA\ U^{\varkappa,\ell,\eta}_{SU(1,1)}(A)\ f(A), \quad \ell\in\{-\tfrac{1+\varkappa}{2}+i[0,+\infty)\}\cup\{0,1,\ldots\},$$

$$\int_{SU(1,1)} dA\,|f(A)|^2 = \sum_{\varkappa=0,1}\Big\{\frac{1}{i\pi}\int_{-(1+\varkappa)/2}^{-(1+\varkappa)/2+i\infty} d\ell\ \ (2\ell+\varkappa+1)\,\pi\,\mathrm{ctg}\,\pi\ell\ \times$$

$$\times\ \mathrm{Sp}(\hat{f}^{\varkappa,\ell,o\dagger}\hat{f}^{\varkappa,\ell,o}) +$$

$$+ \sum_{\ell=0}^{\infty}(2\ell+\varkappa+1)\sum_{\eta=\pm}\mathrm{Sp}(\hat{f}^{\varkappa,\ell,\eta\dagger}\hat{f}^{\varkappa,\ell,\eta})\Big\}.$$

An abstract formulation of these relations is

$$f(A) = \int_{\hat{G}} d\hat{\mu}(\rho)\ \mathrm{Sp}(\hat{f}^{\rho}\ U^{\rho}_{G}(A)^{\dagger}), \qquad A\in G,$$

$$(2.6.77) \qquad \hat{f}^{\rho} = \int_{G} dA\ U^{\rho}_{G}(A)\ f(A), \qquad \rho\in\hat{G},$$

$$\int_{G} dA\,|f(A)|^2 = \int_{\hat{G}} d\hat{\mu}(\rho)\ \mathrm{Sp}(\hat{f}^{\rho\dagger}\hat{f}^{\rho}),$$

where $\hat{G}$ is the set of equivalence classes of irreducible unitary representations of G, $\rho \in \hat{G}$, and $\hat{\mu}$ is the Plancherel measure on $\hat{G}$ defined in the concrete cases by (2.6.76). Obviously $\mathcal{L}^2(\hat{G})$ has the structure

$$(2.6.78) \qquad \mathcal{L}^2(\hat{G}) = \oplus_{\hat{G}}\int \sqrt{d\hat{\mu}(\rho)}\ \mathcal{B}^2(\mathfrak{h}_G^\rho) \quad ,$$

where $\mathcal{B}^2(\mathfrak{h}_G^\rho)$ is the Hilbert space of all Hilbert-Schmidt operators in the representation space $\mathfrak{h}_G^\rho$, i.e. of all bounded linear operators in the representation space $\mathfrak{h}_G^\rho$ with a finite norm with respect to the scalar product defined by

$$(2.6.79) \qquad \langle \hat{f}'^\rho | \hat{f}^\rho \rangle_G^\rho := \mathrm{Sp}(\hat{f}'^{\rho\dagger}\hat{f}^\rho) \ , \ \hat{f}'^\rho, \ \hat{f}^\rho \in \mathcal{B}^2(\mathfrak{h}_G^\rho) \quad .$$

This allows the following interpretation of the generalized Fourier transformation (2.6.77): To each $f \in \mathcal{L}^2(G)$ corresponds an operator field $\hat{f}: \hat{G} \longrightarrow \mathcal{L}^2(\hat{G})$ where $\hat{f}^\rho$ for $\hat{\mu}$-almost every ρ is a Hilbert-Schmidt operator on $\mathfrak{h}_G^\rho$ and $\int_{\hat{G}} d\hat{\mu}(\rho)\ \mathrm{Sp}(\hat{f}^{\rho\dagger}\hat{f}^\rho) < \infty$, while conversely to each such operator field corresponds an element from $\mathcal{L}^2(G)$. This map between $\mathcal{L}^2(G)$ and $\mathcal{L}^2(\hat{G})$ is unitary.

In addition we point to a feature in which the Fourier analysis on compact topological groups differs from that on noncompact ones. While for the Fourier analysis of the square-integrable functions on compact groups all equivalence classes of irreducible unitary representations are needed generally this is not the case for noncompact groups. Especially for SU(1,1) already BARGMANN [13] has pointed out that the representations of the supplementary series are not needed for the expansion of elements from $\mathcal{L}^2(SU(1,1))$. But this does not exclude that these representations play an important role in the functional analysis of more general function spaces on SU(1,1) as for instance $\mathcal{L}^p(SU(1,1))$, $1 \leq p \leq \infty$, (cf. KUNZE, STEIN [28]) or also in connection with certain physical questions (cf. HADJEOANNOU [4]). In Chapter 3, however, we shall use solely the Fourier analysis on $\mathcal{L}^2(G)$.

Finally we give an account of the generalized Fourier analysis on the space $\mathcal{L}^2(E(2))$ of square-integrable functions on the group E(2). The Haar measure on E(2) is defined by

$$(2.6.80) \qquad \int dA = \int \frac{d\ \mathrm{arc}\ z}{2\pi} |z|\, d|z| \frac{d\varphi}{4\pi} \ , \quad A = \begin{pmatrix} e^{i\varphi/2} & ze^{-i\varphi/2} \\ 0 & e^{-i\varphi/2} \end{pmatrix} \quad .$$

Then holds the generalized Fourier expansion

$$f(A) = \sum_{\varkappa=0,1} \int_0^\infty d\rho\, \rho\, \mathrm{Sp}(\hat{f}^{\rho,\varkappa}\, U_{E(2)}^{\rho,\varkappa}(A)^\dagger) \quad , \qquad A \in E(2) \quad ,$$

$$(2.6.81) \qquad \hat{f}^{\rho,\varkappa} = \int_{E(2)} dA\, U_{E(2)}^{\rho,\varkappa}(A)\, f(A) \quad , \qquad \rho \in (0,\infty) \quad ,$$

$$\int_{E(2)} dA\, |f(A)|^2 = \sum_{\varkappa=0,1} \int_0^\infty d\rho\, \rho\, \mathrm{Sp}(\hat{f}^{\rho,\varkappa\dagger} \hat{f}^{\rho,\varkappa}) \quad .$$

It contains essentially the Hankel transformation

$$(2.6.82) \qquad \begin{aligned} f(|z|) &= \int_0^\infty d\rho\, \rho\, \hat{f}(\rho)\, J_{\mu'-\mu}(\rho|z|) \quad , \\ \hat{f}(\rho) &= \int_0^\infty d|z||z|\, J_{\mu'-\mu}(\rho|z|)\, f(|z|) \end{aligned}$$

in $\mathcal{L}^2(0,\infty)$ (cf. TITCHMARSH[29]).

2.7 Expansions of Square-Integrable Functions on the Coset Spaces $SU(2)/H_1$, $SU(1,1)/H_1$ and $E(2)/H_1$

The parametrizations (2.6.11) of the groups SU(2) and SU(1,1) by Eulers angles, the parametrization (1.4.1) of the group E(2) and the Iwasawa decomposition (2.5.57) of SU(1,1) allows us to choose as representants of the cosets of these groups with respect to the subgroups H_1 and - in the case of SU(1,1) - also the subgroup H_2 the following elements:

$$A(i\beta,\varphi): \qquad SU(2)/H_1 = \{ A(i\beta,\varphi)H_1 : 0 \le \varphi < 2\pi,\ 0 \le \beta \le \pi \} ,$$

$$A(\xi,\varphi): \qquad SU(1,1)/H_1 = \{ A(\xi,\varphi)H_1 : 0 \le \varphi < 2\pi,\ 0 \le \xi < \infty \} ,$$

$$A'(x,\varphi): \qquad SU(1,1)/H_2 = \{ A'(x,\varphi)H_2 : 0 \le \varphi < 2\pi,\ -\infty < x < \infty \} ,$$

$$(2.7.1) \qquad A(z) \quad : \qquad E(2)/H_1 = \{ A(z)H_1 : z \in \mathbb{C} \} ;$$

$$A(\xi,\varphi) := \begin{pmatrix} \cosh\xi/2 & e^{i\varphi}\sinh\xi/2 \\ e^{-i\varphi}\sinh\xi/2 & \cosh\xi/2 \end{pmatrix} ,$$

$$A'(x,\ ) := \begin{pmatrix} e^{i\varphi/2}(1-ix/2) & -e^{i\varphi/2}ix/2 \\ e^{-i\varphi/2}ix/2 & e^{-i\varphi/2}(1+ix/2) \end{pmatrix} ,$$

$$A(z) := \begin{pmatrix} 1 & z \\ 0 & 1 \end{pmatrix} .$$

Any $A \in G$, $G \in \{SU(2), SU(1,1), E(2)\}$ therefore has the unique decomposition $A = R(q)Q$ with $Q \in H$, $H \in \{H_1, H_2\}$ and $R(q)$ as a representant of a coset from G/H of the form given by (2.7.1). Here q denotes a parameter characterizing the cosets that will be given a concrete form not before (3.1.18). The Haar measures (2.6.68) on SU(2) and SU(1,1) or (2.6.80) on E(2) in the parametrization (2.7.1) are written

$$\text{(2.7.2)} \quad \int dA = \int dR(q) \int dQ = \begin{cases} \int \frac{d\varphi}{2\pi}\, d(-\sinh^2 \xi/2) \int \frac{d\alpha}{4\pi} & \text{for } A = A(\xi,\varphi)C(\alpha)\ , \\ \int \frac{d\varphi}{2\pi} \frac{dx}{2}\ \ \frac{1}{2}\sum_{\varepsilon} \int \frac{d\xi}{2\pi} & \text{for } A = A'(x,\varphi)D(\varepsilon,\xi), \\ \int \frac{d \arg z}{2}\, |z|\, d|z| \int \frac{d\alpha}{4\pi} & \text{for } A = A(z)C(\alpha)\ . \end{cases}$$

This defines a decomposition of the Haar measure on G into the product of a quasi-invariant - here even invariant - measure on G/H and the Haar measure on H. We now derive a variant of the Fourier expansion of Section 2.6 needed in Chapter 3. In the case of the compact subgroup $H = H_1$ one obtains from this complete orthonormal bases in $\mathcal{L}^2(G/H_1)$.

Let σ be a parameter running through the set $\hat{H}$ of all equivalence classes of irreducible unitary representations of H, $\hat{\nu}$ the Plancherel measure on $\hat{H}$ defined by (2.1.4) and (2.1.9), respectively, τ the index of a basis in the Hilbert space component $\mathfrak{h}_G^{\rho,\sigma}$ of $\breve{A}\mathfrak{h}_G^{\rho}$ on which the representation $\breve{A}\, U_G^{\rho}|H\, \breve{A}^{-1}$ operates as a multiple of the irreducible unitary representation χ^{σ} of H, $\hat{H}_{\rho}$ the set of the $\sigma \in \hat{H}$ for which χ^{σ} appears in the reduction of $U_G^{\rho}|H$. Using the generalized matrix elements $U_G^{\rho}(A)_{\tau'\sigma',\tau\sigma}$ in a basis belonging to H we write the expansion formulas (2.6.77) for elements from $\mathcal{L}^2(G)$ in the form

$$\text{(2.7.3)} \quad \begin{aligned} f(A) &= \int_{\hat{G}} d\hat{\mu}(\rho) \sum_{\tau',\tau} \int_{\hat{H}_{\rho}\times\hat{H}_{\rho}} d\hat{\nu}(\sigma')d\hat{\nu}(\sigma)\ \hat{f}^{\rho}_{\tau'\sigma',\tau\sigma}\ U_G^{\rho}(A)^{*}_{\tau'\sigma',\tau\sigma}\ , \\ \hat{f}^{\rho}_{\tau'\sigma',\tau\sigma} &= \int_G dA\ U_G^{\rho}(A)_{\tau'\sigma',\tau\sigma}\ f(A) \qquad . \end{aligned}$$

Here we use the fact that for the measure $\tilde{\nu}_{\rho}$ on $\hat{H}$, defined by the decomposition of $U_G^{\rho}|H$ which was mentioned in the introduction to this Chapter, in all cases holds the relation

$$\text{(2.7.4)} \qquad \int_{\hat{H}} d\tilde{\nu}_{\rho}(\sigma) = \int_{\hat{H}_{\rho}} d\hat{\nu}(\sigma) \qquad .$$

This follows from Sections 2.2 - 2.5. Since $f \in \mathcal{L}^2(G)$, the function on the subgroup H defined by f(A) = f(R(q)Q) for almost every R(q) lies in $\mathcal{L}^2(H)$. For almost every R(q) therefore the Fourier analysis on $\mathcal{L}^2(H)$, given by (2.1.3), (2.1.4) and (2.1.8), (2.1.9), yields the expansion

$$f(R(q)Q) = \int_{\hat{H}} d\hat{\nu}(\sigma)\, f_\sigma(R(q))\, \chi^\sigma(Q)^* \,,$$

$$(2.7.5) \qquad f_\sigma(R(q)) = \int_H dQ\, \chi^\sigma(Q)\, f(R(q)Q) \,,$$

$$\int_H dQ\, |f(R(q)Q)|^2 = \int_{\hat{H}} d\hat{\nu}(\sigma)\, |f_\sigma(R(q))|^2 \,.$$

With this and the relation

$$(2.7.6) \qquad U_G^\rho(R(q)Q)_{\tau'\sigma',\tau\sigma} = U_G^\rho(R(q))_{\tau'\sigma',\tau\sigma}\, \chi^\sigma(Q) \,,$$

holding for the generalized matrix elements, we get from (2.7.3) the expansion

$$(2.7.7) \qquad \begin{aligned} f_\sigma(R(q)) &= \int_{\hat{G}_\sigma} d\hat{\mu}(\rho) \int_{\hat{H}_\rho} d\hat{\nu}(\sigma') \sum_{\tau',\tau} \hat{f}^\rho_{\tau'\sigma',\tau\sigma}\, U_G^\rho(R(q))^*_{\tau'\sigma',\tau\sigma} \,, \\ \hat{f}^\rho_{\tau'\sigma',\tau\sigma} &= \int_{G/H} dR(q)\, U_G^\rho(R(q))_{\tau'\sigma',\tau\sigma}\, f_\sigma(R(q)) \,. \end{aligned}$$

Here $\hat{G}_\sigma$ denotes the set of all $\rho \in \hat{G}$ for which $U_G^\rho|H$ contains a multiple of χ^σ. The unitarity condition (2.6.77) written in the component form of (2.7.3) reads

$$(2.7.8) \qquad \int_G dA\, |f(A)|^2 = \int_{\hat{G}} d\hat{\mu}(\rho) \int_{\hat{H}_\rho} d\hat{\nu}(\sigma') \int_{\hat{H}_\rho} d\hat{\nu}(\sigma) \sum_{\tau',\tau} |\hat{f}^\rho_{\tau'\sigma',\tau\sigma}|^2 \,.$$

With (2.7.5) we arrive at

$$(2.7.9) \qquad \int_{\hat{H}} d\hat{\nu}(\sigma) \int_{G/H} dR(q)\, |f_\sigma(R(q))|^2 = \int_{\hat{H}} d\hat{\nu}(\sigma) \int_{\hat{G}_\sigma} d\hat{\mu}(\rho) \int_{\hat{H}_\rho} d\hat{\nu}(\sigma') \sum_{\tau',\tau} |\hat{f}^\rho_{\tau'\sigma',\tau\sigma}|^2 \,,$$

i.e. the Fourier expansion (2.7.7) can be read as a unitary map between the Hilbert spaces $\mathcal{L}^2(\hat{H}) \otimes \mathcal{L}^2(G/H) = \oplus \int_{\hat{H}} \sqrt{d\hat{\nu}(\sigma)}\, \mathcal{L}^2(G/H)$ and $\mathcal{L}^2(\hat{G})$ with the structure

$$(2.7.10) \qquad \begin{aligned} \mathcal{L}^2(\hat{G}) &= \oplus \int_{\hat{H}} \sqrt{d\hat{\nu}(\sigma)}\, \mathcal{L}^2(\hat{G})^\sigma \,, \\ \mathcal{L}^2(G)^\sigma &:= \oplus \int_{\hat{G}_\sigma} \sqrt{d\hat{\mu}(\rho)} \left[\oplus \int_{\hat{H}_\rho} \sqrt{d\hat{\nu}(\sigma')} \left(\oplus \sum_{\tau',\tau} \mathbb{C} \right) \right]. \end{aligned}$$

In the case of the compact subgroup $H = H_1$ from (2.7.7) further conclusions may be deduced. Here $\hat{H} = \hat{H}_1$ is the discrete set of all pairs $\sigma = (\varkappa,\mu)$, $\varkappa \in \{0,1\}$, $\mu \in \{0,\pm 1,\pm 2, \dots\}$, and $\int d\hat{\nu}(\sigma)$ is the sum $\sum_{\varkappa,\mu}$; the indices τ',τ are not present. By the Fourier expansion (2.7.5) on H_1 $\mathcal{L}^2(G)$ is decomposed into the direct sum

(2.7.11)
$$\mathcal{L}^2(G) = \oplus\sum_{\varkappa=0,1} \oplus\sum_{\mu=-\infty}^{+\infty} \mathcal{L}^2(G)^{\varkappa,\mu} ,$$
$$\mathcal{L}^2(G)^{\varkappa,\mu} := \{f \in \mathcal{L}^2(G) : f(R(q)Q) = f(R(q))\, \chi^{\varkappa,\mu}(Q)^*,\ Q \in H\}.$$

Take $f \in \mathcal{L}^2(G)^{\varkappa,\mu}$. Then because of (2.7.5)

(2.7.12)
$$f_{\varkappa',\mu'}(R(q)) = \delta_{\varkappa'\varkappa}\delta_{\mu'\mu}\, f(R(q))$$

and because of (2.7.7)

(2.7.13)
$$\hat{f}^{\rho}_{\varkappa''\mu'',\varkappa'\mu'} = \delta_{\varkappa'\varkappa}\delta_{\mu'\mu} \int_{G/H} dR(q)\ U^{\rho}_G(R(q))_{\varkappa''\mu'',\varkappa\mu}\, f(R(q)) .$$

If we decompose $\mathcal{L}^2(\hat{G})$ according to (2.7.10) in the form

(2.7.14)
$$\mathcal{L}^2(\hat{G}) = \oplus\sum_{\varkappa=0,1} \oplus\sum_{\mu=-\infty}^{+\infty} \mathcal{L}^2(\hat{G})^{\varkappa,\mu} ,$$
$$\mathcal{L}^2(\hat{G})^{\varkappa,\mu} := \oplus\int_{\hat{G}_{\varkappa,\mu}} \sqrt{d\hat{\mu}(\rho)}\ \oplus\sum_{(\varkappa,\mu')\in \hat{H}_{1,\rho}} \mathbb{C} ,$$

it follows that in the unitary transformation (2.7.7) between $\mathcal{L}^2(G)$ and $\mathcal{L}^2(\hat{G})$ exactly the subspaces $\mathcal{L}^2(G)^{\varkappa,\mu}$ and $\mathcal{L}^2(\hat{G})^{\varkappa,\mu}$ are mapped unitarily on each other. Since obviously $\mathcal{L}^2(G)^{\varkappa,\mu}$ for every $(\varkappa,\mu)$ is naturally isomorphic to $\mathcal{L}^2(G/H_1)$ we can write instead of (2.7.7) and (2.7.9)

(2.7.15)
$$f(q) = \int_{\hat{G}_{\varkappa,\mu}} d\hat{\mu}(\rho) \sum_{(\varkappa,\mu')\in \hat{H}_{1,\rho}} \hat{f}^{\rho}_{\mu'\mu}\ U^{\rho}_G(R(q))^*_{\mu'\mu} ,$$
$$\hat{f}^{\rho}_{\mu'\mu} = \int_{G/H} dR(q)\ U^{\rho}_G(R(q))_{\mu'\mu}\, f(q) ,$$
$$\int_{G/H} dR(q)\ |f(q)|^2 = \int_{\hat{G}_{\varkappa,\mu}} d\hat{\mu}(\rho) \sum_{(\varkappa,\mu')\in \hat{H}_{1,\rho}} |\hat{f}^{\rho}_{\mu'\mu}|^2 .$$

Here $\hat{f}^{\rho}_{\mu'\mu}$ and $U^{\rho}_G(R(q))_{\mu'\mu}$ are defined by

(2.7.16)
$$\hat{f}^{\rho}_{\varkappa''\mu',\varkappa'\mu} = \delta_{\varkappa''\varkappa}\delta_{\varkappa'\varkappa}\hat{f}^{\rho}_{\mu'\mu},\quad U^{\rho}_G(A)_{\varkappa''\mu',\varkappa'\mu} = \delta_{\varkappa''\varkappa}\delta_{\varkappa'\varkappa} U^{\rho}_G(A)_{\mu'\mu} \quad \text{for } \rho \in \hat{G}_{\varkappa,\mu} .$$

The functions $U_G^\rho(A)_{\mu'\mu}$ obviously are the matrix elements in the H_1-basis calculated in Sections 2.2 - 2.4. To each value of the pair $(\varkappa,\mu)$ belongs a complete orthonormal basis in $\mathcal{L}^2(G/H_1)$:

$$\{{}_\mu Y_{\mu'}^\rho : \rho \in \hat{G}_{\varkappa,\mu},\ (\varkappa,\mu')\in \hat{H}_{1,\rho},\ {}_\mu Y_{\mu'}^\rho(q) = U_G^\rho(R(q))_{\mu'\mu}\} \quad . \tag{2.7.17}$$

$\hat{G}_{\varkappa,\mu}$ and $\hat{H}_{1,\rho}$ in the different cases are given by

<u>G = SU(2):</u>

$$\hat{G}_{\varkappa,\mu} = \{(\varkappa',\ell) : \varkappa' = \varkappa,\ \ell \geq |\mu+\varkappa/2| - \varkappa/2\} \ ,$$

$$\hat{H}_{1,(\varkappa,\ell)} = \{(\varkappa',\mu') : \varkappa' = \varkappa,\ -\ell-\varkappa \leq \mu' \leq \ell\} \ ;$$

<u>G = SU(1,1):</u>

$$\begin{aligned} \hat{G}_{\varkappa,\mu} &= \{(\varkappa',\ell,0) : \varkappa' = \varkappa,\ \ell = -(1+\varkappa)/2 + i\xi,\ \xi \geq 0\} \cup \\ &\cup \{(\varkappa',\ell,\eta) : \varkappa' = \varkappa,\ 0 \leq \ell \leq |\mu+\varkappa/2| - \varkappa/2 - 1,\ \eta = \mathrm{sign}(\mu+\varkappa/2)\} \ , \end{aligned} \tag{2.7.18}$$

$$\hat{H}_{1,(\varkappa,\ell,0)} = \{(\varkappa',\mu') : \varkappa' = \varkappa,\ -\infty < \mu' < \infty\} \ ,$$

$$\hat{H}_{1,(\varkappa,\ell,\pm)} = \{(\varkappa',\mu') : \varkappa' = \varkappa,\ \mathrm{sign}(\mu'+\varkappa/2) = \pm,\ |\mu'+\varkappa/2| - \varkappa/2 \geq \ell+1\} \ ;$$

<u>G = E(2):</u>

$$\hat{G}_{\varkappa,\mu} = \{(\rho,\varkappa') : \varkappa' = \varkappa,\ \rho > 0\} \ ,$$

$$\hat{H}_{1,(\rho,\varkappa)} = \{(\varkappa',\mu') : \varkappa' = \varkappa,\ -\infty < \mu' < \infty\} \ .$$

Since $SU(2)/H_1$ according to (2.7.1) is homeomorphic with the two-dimensional sphere, by (2.7.17) for each value of $\varkappa$ and μ a generalized system of spherical harmonics is defined. Especially for $\varkappa = \mu = 0$ we get, up to normalization factors, the ordinary spherical harmonics $\{e^{i\mu'\varphi} P_\ell^{\mu'}(\cos\beta) : 0 \leq \ell,\ -\ell \leq \mu' \leq \ell\}$. According to (2.7.1) $SU(1,1)/H_1$ is homeomorphic with one shell of a two shell hyperboloid in $\mathbb{R}^3$. The corresponding functions ${}_\mu Y_{\mu'}^{\varkappa,\ell,\eta}$ from (2.7.17) in this case may be interpreted as integral kernels of a Fourier transformation on this surface. From mathematical literature we know the case $\varkappa = \mu = 0$ in which one gets Mehlers conical functions $\{e^{i\mu'\varphi}\, P^{\mu'}_{-1/2+ip}(\cosh\xi):\ p \geq 0,\ -\infty < \mu' < \infty\}$. In the case $E(2)/H_1$ one gets the system $\{{}_{\mu}Y^{\rho,\varkappa}_{\mu'}(z) = e^{i(\mu'-\mu)\operatorname{arc} z}\, J_{\mu'-\mu}(\rho|z|):\ \rho > 0,\ -\infty < \mu' < \infty\}$ with the Bessel functions $J_{\mu'-\mu}$ for the Fourier expansion on the euclidean plane in polar coordinates.

In the case of the noncompact subgroup $H = H_2$ we have $\hat{H} = \{(\varkappa,\lambda):\ \varkappa = 0,1;\ -\infty < \lambda < \infty\}$ and $\int d\hat{\nu}(\mathfrak{G}) = \sum_{\varkappa} \int d\lambda$; the indices τ',τ take the values $\pm$. A Fourier analysis which is analogous to the expansion in $\mathcal{L}^2(G/H_1)$ cannot be formulated here. The unitary transformation (2.7.7) between $\oplus_{\hat{H}}\int\sqrt{d\hat{\nu}(\mathfrak{G})}\ \mathcal{L}^2(G/H)$ and $\oplus_{\hat{H}}\int\sqrt{d\hat{\nu}(\mathfrak{G})}\ \mathcal{L}^2(\hat{G})^{\mathfrak{G}}$ is established by the generalized integral kernels

$$\left\{ {}_{\lambda}Y^{\rho}_{\lambda'\tau'\tau}:\ \rho \in \hat{G}_{\varkappa,\lambda},\ (\varkappa,\lambda') \in \hat{H}_{2,\rho},\ \tau',\tau = \pm,\quad {}_{\lambda}Y^{\rho}_{\lambda'\tau'\tau}(q) = U^{\rho}_{SU(1,1)}(R(q))_{\tau'\lambda',\tau\lambda} \right\}. \tag{2.7.19}$$

Here we have

$$\begin{aligned} \hat{G}_{\varkappa,\lambda} &= \{(\varkappa',\ell,0):\ \varkappa' = \varkappa,\ \ell = -(1+\varkappa)/2 + i\xi,\ \xi \geq 0\} \cup \\ &\cup \{(\varkappa',\ell,+):\ \varkappa' = \varkappa,\ \ell = 0,1,\ldots\} \cup \{(\varkappa',\ell,-):\ \varkappa' = \varkappa,\ \ell = 0,1,\ldots\}, \\ \hat{H}_{2,(\varkappa,\ell,\eta)} &= \{(\varkappa',\lambda'):\ \varkappa' = \varkappa,\ -\infty < \lambda' < \infty\}. \end{aligned} \tag{2.7.20}$$

We recall that according to (2.5.74) for the discrete series ${}_{\lambda}Y^{\varkappa,\ell,\eta}_{\lambda'\tau'\tau}$ is different from zero only when $\tau' = \tau = \eta$.

3. The Reduction of the Product of Two Irreducible Unitary Representations of $\tilde{P}$

The direct product $U^{\mathring{p}_1,\rho_1} \otimes U^{\mathring{p}_2,\rho_2}$ of two irreducible unitary representations $U^{\mathring{p}_1,\rho_1}$ and $U^{\mathring{p}_2,\rho_2}$ of $\tilde{P}$ is defined as the representation

$$
\begin{aligned}
&((U^{\mathring{p}_1,\rho_1} \otimes U^{\mathring{p}_2,\rho_2})(A,a)\psi)(p_1,p_2) \equiv (U^{1,2}(A,a)\psi)(p_1,p_2) = \\
(3.1)\quad &= U^{\mathring{p}_1,\rho_1}_{G(\mathring{p}_1)}(R(p_1;A),\Lambda(p_1)^{-1}a)\otimes U^{\mathring{p}_2,\rho_2}_{G(\mathring{p}_2)}(R(p_2;A),\Lambda(p_2)^{-1}a)\,\psi(\Lambda^{-1}p_1,\Lambda^{-1}p_2)
\end{aligned}
$$

with $\Lambda = \Lambda(A)$ on the Hilbert space

$$
\begin{aligned}
&\mathfrak{h}^{1,2} := \bigoplus\!\!\int_{\Omega(\mathring{p}_1)\times\Omega(\mathring{p}_2)} \sqrt{d\omega_{\mathring{p}_1}(p_1)d\omega_{\mathring{p}_2}(p_2)}\; \mathfrak{h}^{\rho_1}_{G(\mathring{p}_1)}(p_1) \otimes \mathfrak{h}^{\rho_2}_{G(\mathring{p}_2)}(p_2)\,, \\
(3.2)\quad &\mathfrak{h}^{\rho_i}_{G(\mathring{p}_i)}(p_i) \equiv \mathfrak{h}^{\rho_i}_{G_i},\quad G_i := G(\mathring{p}_i)\,,\ i = 1,2\,,
\end{aligned}
$$

of vector valued functions $\psi: \Omega(\mathring{p}_1)\times\Omega(\mathring{p}_2) \longrightarrow \mathfrak{h}^{\rho_1}_{G_1}\otimes\mathfrak{h}^{\rho_2}_{G_2}$ with scalar product

$$
(3.3)\quad \langle\psi|\varphi\rangle^{1,2} := \int_{\Omega(\mathring{p}_1)\times\Omega(\mathring{p}_2)} d\omega_{\mathring{p}_1}(p_1)d\omega_{\mathring{p}_2}(p_2)\, \langle\psi(p_1,p_2)|\,\varphi(p_1,p_2)\rangle^{\rho_1,\rho_2}\,.
$$

Here $\langle|\rangle^{\rho_1,\rho_2}$ means the scalar product in $\mathfrak{h}^{\rho_1}_{G_1}\otimes\mathfrak{h}^{\rho_2}_{G_2}$. According to (1.1.15) we have

$$
\begin{aligned}
&U^{\mathring{p}_1,\rho_1}_{G(\mathring{p}_1)}(R(p_1;A),\Lambda(p_1)^{-1}a) \otimes U^{\mathring{p}_2,\rho_2}_{G(\mathring{p}_2)}(R(p_2;A),\Lambda(p_2)^{-1}a) = \\
(3.4)\quad &= e^{i(p_1+p_2)\cdot a}\, U^{\rho_1}_{G_1}(R(p_1;A)) \otimes U^{\rho_2}_{G_2}(R(p_2;A))\,.
\end{aligned}
$$

For the product representations $U^{1,2}$ with nonvanishing momenta $\mathring{p}_1 \neq 0$, $\mathring{p}_2 \neq 0$ we construct in Section 3.1 the unitary operator that transforms $U_{1,2}$ into a direct integral of multiples of irreducible unitary representations of $\tilde{P}$. The irreducible unitary representations occurring in this decomposition, their multiplicities as well as the Clebsch-Gordan coefficients leading to the decomposition are listed in detail in Section 3.2. The decomposition of the product representation $U^{1,2}$ with at least one factor with vanishing momentum is sketched in Section 3.3.

3.1 The Decomposition of the Product Representation $U^{\mathring{p}_1,\rho_1} \otimes U^{\mathring{p}_2,\rho_2}$ with $\mathring{p}_1 \neq 0 \neq \mathring{p}_2$ into Irreducible Representations

To the representation $U^{1,2}$ according to (3.1) and (3.4) belongs the character

$$\chi^p(a) = e^{ip\cdot a},\ p = p_1 + p_2 \tag{3.1.1}$$

on the translation subgroup $\check{\mathbb{R}}^4 = (\mathbb{1}_2, \mathbb{R}^4)$ of $\tilde{P}$. In $U^{1,2}$ therefore occur at most such representations $U^{\mathring{p},\rho}$ for which a $p_1 \in \Omega(\mathring{p}_1)$ and a $p_2 \in \Omega(\mathring{p}_2)$ exist such that the total momentum $p = p_1 + p_2$ is from $\Omega(\mathring{p})$. The following Table shows the mass shells $\Omega(\mathring{p})$ on which the total momentum p varies if p_1 and p_2 run through their mass shells $\Omega(\mathring{p}_1)$ and $\Omega(\mathring{p}_2)$. To the pairs $(\mathring{p}_1, \mathring{p}_2)$ correspond the partial domains $\Omega_k \subset \Omega$ (cf. (1.1.2)) in which the standard momentum $\mathring{p}$ varies.

Table 3.1

I	$\mathring{p}_1 = m_1 e_{(0)},\ \mathring{p}_2 = m_2 e_{(0)};$	$\mathring{p} \in \Omega_I$	$:= \{me_{(0)}:\ m \geq m_1 + m_2\}$,
II	$\mathring{p}_1 = m_1 e_{(0)},\ \mathring{p}_2 = e_{(0)} + e_{(3)};$	$\mathring{p} \in \Omega_{II}$	$:= \{me_{(0)}:\ m > m_1\}$,
III	$\mathring{p}_1 = m_1 e_{(0)},\ \mathring{p}_2 = n_2 e_{(3)};$	$\mathring{p} \in \Omega_{III}$	$:= \{me_{(0)}:\ m > 0\} \cup \{ne_{(3)}:\ n > 0\} \cup \{e_{(0)} + e_{(3)}\}$,
IV	$\mathring{p}_1 = m_1 e_{(0)},\ \mathring{p}_2 = -e_{(0)} - e_{(3)};$	$\mathring{p} \in \Omega_{IV}$	$:= \{me_{(0)}:\ 0 < m < m_1\} \cup \{ne_{(3)}:\ n > 0\} \cup \{e_{(0)} + e_{(3)}\}$,
V	$\mathring{p}_1 = m_1 e_{(0)},\ \mathring{p}_2 = -m_2 e_{(0)};$	$\mathring{p} \in \Omega_V$	$:= \{me_{(0)}:\ 0 < m \leq m_1 - m_2\} \cup \{e_{(0)} + e_{(3)}\} \cup \{ne_{(3)}:\ n > 0\}$ for $m_1 > m_2$,
			$= \{ne_{(3)}:\ n > 0\} \cup \{0\}$ for $m_1 = m_2$,
VI	$\mathring{p}_1 = e_{(0)} + e_{(3)},\ \mathring{p}_2 = e_{(0)} + e_{(3)};$	$\mathring{p} \in \Omega_{VI}$	$:= \{me_{(0)}:\ m > 0\} \cup \{e_{(0)} + e_{(3)}\}$,

Table 3.1 (continued)

VII : $\mathring{p}_1 = e_{(0)}+e_{(3)}$, $\mathring{p}_2 = n_2 e_{(3)}$; $\mathring{p} \in \Omega_{VII}$: $= \{me_{(0)}: m > 0\} \cup \{e_{(0)}+e_{(3)}\} \cup$
$\cup \{ne_{(3)}: n > 0\}$,

VIII: $\mathring{p}_1 = e_{(0)}+e_{(3)}$, $\mathring{p}_2 = -e_{(0)}-e_{(3)}$; $\mathring{p} \in \Omega_{VIII}$: $= \{ne_{(3)}: n > 0\} \cup \{0\}$,

IX : $\mathring{p}_1 = n_1 e_{(3)}$, $\mathring{p}_2 = n_2 e_{(3)}$; $\mathring{p} \in \Omega_{IX}$: $= \{me_{(0)}: m > 0\} \cup \{-me_{(0)}: m > 0\} \cup$
$\cup \{e_{(0)}+e_{(3)}\} \cup \{-e_{(0)}-e_{(3)}\} \cup$
$\cup \{ne_{(3)}: n > 0\}$ for $n_1 \neq n_2$,
$= \{me_{(0)}: m > 0\} \cup \{-me_{(0)}: m > 0\} \cup$
$\cup \{e_{(0)}+e_{(3)}\} \cup \{-e_{(0)}-e_{(3)}\} \cup$
$\cup \{ne_{(3)}: n > 0\} \cup \{0\}$ for $n_1 = n_2$,

X : $\mathring{p}_1 \neq 0$, $\mathring{p}_2 = 0$; $\mathring{p} \in \Omega_X$: $= \{\mathring{p}_1\}$,

XI : $\mathring{p}_1 = 0 = \mathring{p}_2$; $\mathring{p} \in \Omega_{XI}$: $= \{0\}$.

The cases not listed in this Table are easily obtained by symmetry considerations. For instance the case $m_2 > m_1$ in V is obtained by exchanging the indices 1 and 2 and substituting p by -p, for the little groups belonging to p and -p are identical. Only the cases I through IX are dealt with in detail in the following. For the cases X and XI the construction leading to the decomposition of the product representation $U^{1,2}$ is sketched in Section 3.3.

To a fixed $p \in \Omega(\mathring{p})$, $\mathring{p} \in \Omega_k$, $k \in \{I, \ldots, IX\}$ still belongs a two-parameter set of pairs of momenta $(p_1, p_2) \in \Omega(\mathring{p}_1) \times \Omega(\mathring{p}_2)$ with $p_1 + p_2 = p$. We characterize this set by a vector

$$q = \alpha p_1 - (1-\alpha) p_2 \ , \quad \alpha \in \mathbb{R} \ , \tag{3.1.2}$$

which is linearly independent of p and lies in the plane which is spanned by p_1 and p_2. Since α can be choosen arbitrarily and q satisfies the condition

$$(3.1.3) \qquad (p\cdot q)^2 - p^2q^2 = (p_1\cdot p_2)^2 - p_1{}^2p_2{}^2 \quad ,$$

which is independent of α, q really has only two essential degrees of freedom. For the following we introduce the restrictions

$$(3.1.4) \qquad p^2 \neq 0 \neq q^2 \; ,$$

and choose q orthogonal to p:

$$(3.1.5) \qquad q := \frac{1}{p^2}(p_2\cdot p\; p_1 - p_1\cdot p\; p_2) \; , \qquad p\cdot q = 0 \quad .$$

The restrictions (3.1.4) are not essential since the pairs $(p_1,p_2) \in \Omega(\mathring{p}_1) \times \Omega(\mathring{p}_2)$ for which $p^2 = 0$ or $q^2 = 0$ describe at most a manifold of lower dimension. Being of measure zero this manifold plays no role in the direct integral decomposition of the Hilbert space $\mathfrak{H}^{1,2}$.

The discriminant $(p_1\cdot p_2)^2 - p_1{}^2p_2{}^2$ of the quadratic equation $(xp_1 + p_2)^2 = 0$ in x is positive, zero or negative depending on wether the plane spanned by p_1 and p_2 cuts the light cone in two lines, is a tangential plane to it or is totally spacelike, respectively. The second case we neglect for the same reason as above. Then $(p_1\cdot p_2)^2 - p_1{}^2p_2{}^2$ is positive in the cases I through VIII, but indefinite in the case IX, i.e. only in this case, according to (3.1.3), p and q may be spacelike simultaneously. In all other cases q is spacelike if p is timelike and vice versa.

A very familiar situation is given in the case I where both p_1 and p_2 lie in the foreward light cone. Here in the center of mass system q is exactly the momentum of particle 1:

$$(3.1.6) \qquad \Lambda(p)^{-1}p = \mathring{p} = (m,\vec{0}), \quad \Lambda(p)^{-1}q = (0,\overrightarrow{\Lambda(p)^{-1}p_1}) \quad .$$

$\overrightarrow{\Lambda(p)^{-1}p_1}^{\,2} = q^2$ according to (3.1.3) is fixed by m^2, $m_1{}^2$, $m_2{}^2$ so that q essentially describes only the direction of $\overrightarrow{\Lambda(p)^{-1}p_1}$.

Let $(\mathring{p}_1,\mathring{p}_2)$ be of type k, $k \in \{I, \dots ,IX\}$. By (3.1.1) and (3.1.5) $\Omega(\mathring{p}_1) \times \Omega(\mathring{p}_2)$ is mapped into a region of the p-q-space with the following structure: It consists of the union of the mass shells $\Omega(\mathring{p})$ for $\mathring{p} \in \Omega_k$, and to each $p \in \Omega(\mathring{p})$ belongs a two-dimensional surface

(3.1.7) $$\Sigma(p): = \{q: p\cdot q = 0,\ -p^2q^2 = (p_1\cdot p_2)^2 - p_1{}^2p_2{}^2\}$$

in q-space. Let dR(q) denote a surface element on $\Sigma(\mathring{p})$ that is invariant under the operation

(3.1.8) $$G(\mathring{p}) \ni A: q \longrightarrow \Lambda(A)q\ ,\ q \in \Sigma(\mathring{p})$$

of $G(\mathring{p})$ on $\Sigma(\mathring{p})$:

(3.1.9) $$dR(\Lambda(A)q) = dR(q),\quad q \in \Sigma(\mathring{p})\ ,\ A \in G(\mathring{p})\quad .$$

The existence of such surface elements will be guaranteed by giving explicit expressions in (3.1.20). With this we can define a surface element on $\Sigma(p)$ by

(3.1.10) $$dR_p(q): = dR(\Lambda(p)^{-1}q)\ ,\quad q \in \Sigma(p)\quad .$$

Because of (3.1.9) it obeys the invariance condition

(3.1.11) $$dR_{\Lambda p}(\Lambda q) = dR_p(q)\quad .$$

Let ρ be the measure on Ω defined by

(3.1.12) $$\rho(\mathfrak{M}) = \begin{cases} \int_{\mathfrak{M}} \left|\frac{\lambda(-m^2, p_1{}^2, p_2{}^2)}{16\, m^2}\right|^{1/2} dm & \text{for } \mathfrak{M} \subset \{\pm m e_0 : m > 0\} = \Omega^+ \cup \Omega^-, \\ \int_{\mathfrak{M}} \left|\frac{\lambda(n^2, p_1{}^2, p_2{}^2)}{16\, n^2}\right|^{1/2} dn & \text{for } \mathfrak{M} \subset \{n e_3 : n > 0\} = \Omega^0, \\ 0 & \text{for } \mathfrak{M} \subset \{e_{(0)}+e_{(3)}\} \cup \{-e_{(0)}-e_{(3)}\} \cup \{0\} = \\ & = \Omega_0^+ \cup \Omega_0^- \cup \Omega_0^0\ , \end{cases}$$

$$\lambda(x,y,z): = x^2 + y^2 + z^2 - 2xy - 2yz - 2zx\quad .$$

Then with the invariant measure $\omega_{\mathring{p}}$ on $\Omega(\mathring{p})$ we can introduce a measure on the image of $\Omega(\mathring{p}_1) \times \Omega(\mathring{p}_2)$ in the p-q-space which is invariant under SL(2,$\mathbb{C}$) by

(3.1.13) $$d\rho(\mathring{p})\ d\omega_{\mathring{p}}(p)\ dR_p(q) = d\rho(\mathring{p})\ d\omega_{\mathring{p}}(\Lambda p)\ dR_{\Lambda p}(\Lambda q)\quad .$$

Fixing the normalization of $dR_p(q)$ we then put

$$d\rho(\mathring{p})\; d\omega_{\mathring{p}}(p)\; dR_p(q) = d\omega_{\mathring{p}_1}(p_1)\; d\omega_{\mathring{p}_2}(p_2) \quad . \tag{3.1.14}$$

It is easily seen that the operation (3.1.8) of $G(\mathring{p})$ on $\Sigma(\mathring{p})$ is transitive so that $\Sigma(\mathring{p})$ may be characterized by a standard element $\mathring{q}$ for which $\Sigma(\mathring{p}) = G(\mathring{p})\mathring{q}$. We choose these standard elements according to the following scheme

Table 3.2

I-VIII: $\mathring{p} \in \Omega^{\pm} \longrightarrow \mathring{q} \in \Omega^{0} = \{ne_{(3)}:\ n > 0\}$,

$\mathring{p} \in \Omega^{0} \longrightarrow \mathring{q} \in \Omega^{\pm} = \{\pm m e_{(0)}:\ m > 0\}$,

IX : $\mathring{p} \in \Omega^{\pm} \longrightarrow \mathring{q} \in \Omega^{0}$,

$$\mathring{p} \in \Omega^{0} \longrightarrow \mathring{q} \in \begin{cases} \Omega^{\pm} & \text{for } (p_1\cdot p_2)^2 - p_1^{\,2} p_2^{\,2} > 0 \ , \\ \Omega^{0\prime} := \{ne_{(2)}:\ n > 0\} & \text{for } (p_1\cdot p_2)^2 - p_1^{\,2} p_2^{\,2} < 0 \ . \end{cases}$$

The stability subgroup $G(\mathring{p},\mathring{q}) \subset G(\mathring{p})$ that leaves $\mathring{q}$ invariant has the form

$$G(\mathring{p},\mathring{q}) = G(\mathring{p}) \cap G(\mathring{q}) \ . \tag{3.1.15}$$

According to Table 3.2 there exist only two different such groups, the subgroups

$$\begin{aligned} H_1 &= G(e_{(0)},e_{(3)}) = \left\{ C(\varphi) := \begin{pmatrix} e^{i\varphi/2} & 0 \\ 0 & e^{-i\varphi/2} \end{pmatrix} :\ 0 \le \varphi < 4\pi \right\}, \\ H_2 &= G(e_{(3)},e_{(2)}) = \left\{ D(\varepsilon,\xi) := \varepsilon \begin{pmatrix} \cosh\xi/2 & \sinh\xi/2 \\ \sinh\xi/2 & \cosh\xi/2 \end{pmatrix} :\ \begin{matrix} \varepsilon = \pm\ , \\ -\infty < \xi < \infty \end{matrix} \right\}, \end{aligned} \tag{3.1.16}$$

introduced in Section 2.1. For $\mathring{p} \in \Omega^{0}$, $\mathring{q} \in \Omega^{0\prime}$ we have $G(\mathring{p},\mathring{q}) = H_2$, in all other cases $G(\mathring{p},\mathring{q}) = H_1$. Since $\Sigma(\mathring{p})$ is homeomorphic with the coset space $G(\mathring{p})/G(\mathring{p},\mathring{q})$ we can associate with each $q \in \Sigma(\mathring{p})$ uniquely a representant $R(q) \in G(\mathring{p})$ of a coset from $G(\mathring{p})/G(\mathring{p},\mathring{q})$ so that

$$q = \Lambda(R(q))\mathring{q} \quad . \tag{3.1.17}$$

If we parametrize $\Sigma(\mathring{p})$ by

$$
(3.1.18)\quad q = \begin{cases} \sqrt{q^2}\left[\sin\beta(\sin\varphi\, e_{(1)} + \cos\varphi\, e_{(2)}) + \cos\beta\, e_{(3)}\right], & \\ 0 \le \beta \le \pi,\ 0 \le \varphi < 2\pi, & \text{for } \mathring{p}\in\Omega^{\pm}, \mathring{q}\in\Omega^{0}, \\ \pm\sqrt{-q^2}\left[\cosh\xi\, e_{(0)} + \sinh\xi(\cos\varphi\, e_{(1)} - \sin\varphi\, e_{(2)})\right], & \\ 0 \le \xi < \infty,\ 0 \le \varphi < 2\pi, & \text{for } \mathring{p}\in\Omega^{0}, \mathring{q}\in\Omega^{\pm}, \\ \sqrt{q^2}\left[x\, e_{(0)} + (\sin\varphi - x\cos\varphi)e_{(1)} + (\cos\varphi + x\sin\varphi)e_{(2)}\right], & \\ -\infty < x < \infty,\ 0 \le \varphi < 2\pi, & \text{for } \mathring{p}\in\Omega^{0}, \mathring{q}\in\Omega^{0\prime}, \end{cases}
$$

we can choose the representants according to the following scheme

$$
(3.1.19)\qquad R(q) = \begin{cases} A(i\beta,\varphi) & \text{for } \mathring{p}\in\Omega^{\pm}, \mathring{q}\in\Omega^{0}, \\ A(\xi,\varphi) & \text{for } \mathring{p}\in\Omega^{0}, \mathring{q}\in\Omega^{\pm}, \\ A'(x,\varphi) & \text{for } \mathring{p}\in\Omega^{0}, \mathring{q}\in\Omega^{0\prime}. \end{cases}
$$

Here $A(\xi,\varphi)$ and $A'(x,\varphi)$ are the matrices defined in (2.7.1). The surface element $dR(q)$ of $\Sigma(\mathring{p})$ at the point q according to (2.7.2) with the normalization (3.1.14) has the form

$$
(3.1.20)\qquad dR(q) = \begin{cases} \frac{d\varphi}{2\pi}\, d(\sin^2\beta/2) & \text{for } \mathring{p}\in\Omega^{\pm}, \mathring{q}\in\Omega^{0}, \\ \frac{d\varphi}{2\pi}\, d(-\sinh^2\xi/2) & \text{for } \mathring{p}\in\Omega^{0}, \mathring{q}\in\Omega^{\pm}, \\ \frac{d\varphi}{2\pi}\frac{dx}{2} & \text{for } \mathring{p}\in\Omega^{0}, \mathring{q}\in\Omega^{0\prime}. \end{cases}
$$

With the aid of the matrices $A(p)$ defined in (1.1.9) we now can introduce an operation that carries simultaneously $\mathring{p}$ to p and $\mathring{q}$ to q. With

$$
(3.1.21)\qquad A(p,q) := A(p)R(\Lambda(p)^{-1}q)\ ,\quad \Lambda(p,q) := \Lambda(A(p,q))\ ,
$$

we have

$$
(3.1.22)\qquad \Lambda(p,q)\mathring{p} = p\ ,\quad \Lambda(p,q)\mathring{q} = q\ .
$$

The $A(p,q)$ obviously are representants of the elements of the coset space $SL(2,\mathbb{C})/G(\mathring{p},\mathring{q})$. The momenta p_1 and p_2 are carried by $\Lambda(p,q)^{-1}$ to the system characterized by $\mathring{p}$, $\mathring{q}$. If we choose the abbreviations

(3.1.23) $$\hat{p}_1 := \Lambda(p,q)^{-1}p_1 \ , \quad \hat{p}_2 := \Lambda(p,q)^{-1}p_2$$

for the transformed momenta, we have

(3.1.24) $$\hat{p}_1 + \hat{p}_2 = \mathring{p} \ , \ p_2\cdot p\ \hat{p}_1 - p_1\cdot p\ \hat{p}_2 = p^2\ \mathring{q} \ .$$

In the familiar case $\mathring{p}_1, \mathring{p}_2 \in \Omega^+$ mentioned above the transformation (3.1.23) carries the momenta to the center of mass system and rotates them into the $e_{(3)}$-direction.

For the reduction of the product representation $U^{1,2}$ in (3.1) the result of the reduction of the product representation $U^{\rho_1}_{G_1}(R(p_1;A)) \otimes U^{\rho_2}_{G_2}(R(p_2;A))$ is needed. Here two difficulties occur: First, the groups $G_1 = G(\mathring{p}_1)$ and $G_2 = G(\mathring{p}_2)$ in general are different; secondly, even if $G_1 = G_2$ the arguments $R(p_1;A)$ and $R(p_2;A)$ in general are different. We aim therefore at a unitary transformation of $U^{1,2}$ that gives equal arguments to $U^{\rho_1}_{G_1}$ and $U^{\rho_2}_{G_2}$ which necessarily must lie in the intersection $G_1 \cap G_2$. From (3.1.16) with (1.1.5) we see that in all cases

(3.1.25) $$G(\mathring{p},\mathring{q}) \subset G_1 \cap G_2 \ .$$

In the special cases $G_1 = G_2$ there may exist reduction techniques which are different from the general method applied here. A well known example is the spin-orbit-coupling scheme in the case $\mathring{p}_1, \mathring{p}_2 \in \Omega^+$ (cf. JOOS [1]).

In a next step we pick out from the expression (1.1.15) for $R(p_i;A)$, $i = 1,2$, a common element contained in $G(\mathring{p},\mathring{q})$. For that reason we write $R(p_i;A)$ in the form

(3.1.26)
$$R(p_i;A) = A(p_i)^{-1}A(p,q)Q(p,q;A)A(\Lambda(A)^{-1}p,\Lambda(A)^{-1}q)^{-1}A(\Lambda(A)^{-1}p_i),$$
$$Q(p,q;A) := A(p,q)^{-1}AA(\Lambda(A)^{-1}p,\Lambda(A)^{-1}q) \in G(\mathring{p},\mathring{q}) \ .$$

Now $A(p,q)^{-1}A(p_i)$ carries the standard momentum $\mathring{p}_i$ to the standard system characterized by $\mathring{p}$, $\mathring{q}$, i.e. according to (3.1.23) we have

$$\Lambda(p,q)^{-1}\Lambda(p_i)\mathring{p}_i = \hat{p}_i = \Lambda(\hat{p}_i)\mathring{p}_i \quad . \tag{3.1.27}$$

Therefore

$$R(p_i;A(p,q)) := A(p_i)^{-1}A(p,q)A(\hat{p}_i) \tag{3.1.28}$$

is an element of $G(\mathring{p}_i)$, and we can write for $R(p_i;A)$:

$$\begin{aligned} R(p_i;A) &= R(p_i;A(p,q))A(\hat{p}_i)^{-1}Q(p,q;A)\times \\ &\times A(\hat{p}_i)R(\Lambda(A)^{-1}p_i;A(\Lambda(A)^{-1}p,\Lambda(A)^{-1}q))^{-1} \,. \end{aligned} \tag{3.1.29}$$

Here we have used the relation

$$\widehat{\Lambda p_i} = \hat{p}_i \quad \text{for } \Lambda \in L_+^\uparrow \tag{3.1.30}$$

that is obvious from (3.1.24). Now $Q(p,q;A)$ according to (3.1.16) means either a rotation around the $e_{(3)}$-axis in the $e_{(1)}$-$e_{(2)}$-plane or a velocity transformation in the $e_{(0)}$-$e_{(1)}$-plane. In the first case $\mathring{p}$ and $\mathring{q}$ span the $e_{(0)}$-$e_{(3)}$-plane so that $A(\hat{p}_i)$ according to (1.1.8) and (1.1.9) is a velocity transformation in this plane. In the second case $\mathring{p}$ and $\mathring{q}$ span the $e_{(2)}$-$e_{(3)}$-plane, and $A(\hat{p}_i)$ according to (1.1.8) and (1.1.9) is a rotation in this plane. In both cases therefore $A(\hat{p}_i)$ commutes with $G(\mathring{p},\mathring{q})$.This was intended in choosing the representants $A(p)$ according to (1.1.9). Hence we can bring (3.1.29) to the form

$$R(p_i;A) = R(p_i;A(p,q))Q(p,q;A)R(\Lambda(A)^{-1}p_i;A(\Lambda(A)^{-1}p,\Lambda(A)^{-1}q))^{-1}. \tag{3.1.31}$$

As a first step to the solution of the reduction problem for the product representations we map the representation space $\mathfrak{h}^{1,2}$ by

$$\begin{aligned} &\mathfrak{h}^{1,2} \longrightarrow \mathfrak{h}'^{1,2}\,, \qquad \psi \longmapsto \psi' : \\ &\psi'(p,q) = U_{G_1}^{\rho_1}(R(p_1;A(p,q))^{-1}) \otimes U_{G_2}^{\rho_2}(R(p_2;A(p,q))^{-1})\,\psi(p_1,p_2) \end{aligned} \tag{3.1.32}$$

unitarily onto the Hilbert space

$$\begin{aligned} &\mathfrak{h}'^{1,2} := \oplus\int_{\Omega_k}\sqrt{d\rho(\mathring{p})}\left[\oplus\int_{\Omega(\mathring{p})}\sqrt{d\omega_{\mathring{p}}(p)}\left(\oplus\int_{\Sigma(p)}\sqrt{dR_p(q)}\;\mathfrak{h}^{\rho_1,\rho_2}(p,q)\right)\right] \,, \\ &\mathfrak{h}^{\rho_1,\rho_2}(p,q) \equiv \mathfrak{h}_{G_1}^{\rho_1}\otimes\mathfrak{h}_{G_2}^{\rho_2} \end{aligned} \tag{3.1.33}$$

with scalar product

$$\langle\psi'|\varphi'\rangle'^{1,2} := \int_{\Omega_k} d\rho(\mathring{p}) \int_{\Omega(\mathring{p})} d\omega_{\mathring{p}}(p) \int_{\Sigma(p)} dR_p(q) \langle\psi'(p,q)|\varphi'(p,q)\rangle_{\rho_1\rho_2} \tag{3.1.34}$$

$$= \langle\psi|\varphi\rangle^{1,2} .$$

If the product representation is carried over to $\mathfrak{h}'^{1,2}$ because of (3.1.31) it assumes the form

$$(U'^{1,2}(A,a)\psi')(p,q) := (U^{1,2}(A,a)\psi)'(p,q) =$$

$$= e^{ip\cdot a}\, U^{\rho_1}_{G_1}(Q(p,q;A)) \otimes U^{\rho_2}_{G_2}(Q(p,q;A))\psi'(\Lambda(A)^{-1}p,\Lambda(A)^{-1}q) . \tag{3.1.35}$$

Obviously $U'^{1,2}$ has the form of an induced representation of $\tilde{P}$ with the restricted representation $U^{\mathring{p}_1;\rho_1}_{\check{G}(\mathring{p}_1)} \otimes U^{\mathring{p}_2;\rho_2}_{\check{G}(\mathring{p}_2)}|\check{G}(\mathring{p},\mathring{q})$, $\check{G}(\mathring{p},\mathring{q}) := G(\mathring{p},\mathring{q}) \circledS \mathbb{R}^4 \subset \tilde{P}$ as inducing representation. This, however, in no way is irreducible, for first $U^{\rho_1}_{G_1} \otimes U^{\rho_2}_{G_2}|G(\mathring{p},\mathring{q})$ in general is reducible, and secondly the subgroup $G(\mathring{p},\mathring{q})$ is too small (cf. MACKEY [9]). Both these reasons may serve in the following as a guide to the final solution of the reduction problem.

First we decompose $U^{\rho_1}_{G_1} \otimes U^{\rho_2}_{G_2}|G(\mathring{p},\mathring{q})$ into its irreducible components. In sections 2.2 - 2.5 we have solved the reduction problem for $U^{\rho}_{G}|H$, $H = G(\mathring{p},\mathring{q}) \in \{H_1, H_2\}$, $G \in \{SU(2), SU(1,1), E(2)\}$ (cf. the last section in 2.5). There $\mathfrak{h}^{\rho}_{G}$ was mapped by a unitary equivalence transformation $\check{A}$ onto the Hilbert space

$$\check{A}\,\mathfrak{h}^{\rho}_{G} = \tilde{\mathfrak{h}}^{\rho}_{G,H} := \oplus\int_{\hat{H}_\rho} \sqrt{d\hat{\nu}(\sigma)}\; \mathbb{C}^{n(\rho)} \tag{3.1.36}$$

on which $\tilde{U}^{\rho}_{G}|H$ decomposes into a direct integral of irreducible unitary representations χ^{σ} of H each occurring $n(\rho)$ times:

$$\tilde{U}^{\rho}_{G}(Q)\tilde{f}^{\rho}_{\tau}(\sigma) = \chi^{\sigma}(Q)\tilde{f}^{\rho}_{\tau}(\sigma) \; , \quad \tilde{f}^{\rho} \in \tilde{\mathfrak{h}}^{\rho}_{G,H} . \tag{3.1.37}$$

Let $\tilde{\mathfrak{h}}^{\rho_i}_{G_i,H} = \check{A}_i\,\mathfrak{h}^{\rho_i}_{G_i}$, i = 1,2, be the images of the representation spaces $\mathfrak{h}^{\rho_i}_{G_i}$. Then the direct product $\check{A}_1 \otimes \check{A}_2$ maps $\mathfrak{h}^{\rho_1}_{G_1} \otimes \mathfrak{h}^{\rho_2}_{G_2}$ unitarily onto the Hilbert space

$$(3.1.38)\quad (\ddot{A}_1 \otimes \ddot{A}_2)(\mathfrak{h}^{\rho_1}_{G_1} \otimes \mathfrak{h}^{\rho_2}_{G_2}) = \tilde{\mathfrak{h}}^{\rho_1}_{G_1,H} \otimes \tilde{\mathfrak{h}}^{\rho_2}_{G_2,H} =: \tilde{\mathfrak{h}}^{\rho_1,\rho_2}_{G_1,G_2,H} = \oplus\int_{\hat{H}_{\rho_1}} \sqrt{d\hat{\nu}(\sigma_1)} \oplus \int_{\hat{H}_{\rho_2}} \sqrt{d\hat{\nu}(\sigma_2)}\, (\mathbb{C}^{n(\rho_1)} \otimes \mathbb{C}^{n(\rho_2)})$$

of complex valued functions $\tilde{f} = (\ddot{A}_1 \times \ddot{A}_2)f$, $f \in \mathfrak{h}^{\rho_1}_{G_1} \otimes \mathfrak{h}^{\rho_2}_{G_2}$ on $\{1,\dots,n(\rho_1)\} \times H_{\rho_1} \times \{1,\dots,n(\rho_2)\} \times H_{\rho_2}$ with scalar product

$$(3.1.39)\quad \langle \tilde{f}|\tilde{g}\rangle^{\sim\rho_1,\rho_2} := \int_{\hat{H}_{\rho_1}} d\hat{\nu}(\sigma_1) \int_{\hat{H}_{\rho_2}} d\hat{\nu}(\sigma_2) \sum_{\tau_1=1}^{n(\rho_1)} \sum_{\tau_2=1}^{n(\rho_2)} \tilde{f}_{\tau_1\tau_2}(\sigma_1,\sigma_2)^* \tilde{g}_{\tau_1\tau_2}(\sigma_1,\sigma_2) = \langle f|g\rangle^{\rho_1,\rho_2}.$$

For the transformed representation

$$(3.1.40)\quad \tilde{U}^{\rho_1,\rho_2}_{G_1,G_2} := (\ddot{A}_1 \otimes \ddot{A}_2)(U^{\rho_1}_{G_1} \otimes U^{\rho_2}_{G_2})(\ddot{A}_1 \otimes \ddot{A}_2)^{-1}$$

on $\tilde{\mathfrak{h}}^{\rho_1,\rho_2}_{G_1,G_2,H}$ we have

$$(3.1.41)\quad \tilde{U}^{\rho_1,\rho_2}_{G_1,G_2}(Q)\tilde{f}_{\tau_1\tau_2}(\sigma_1,\sigma_2) = \chi^{\sigma_1}(Q)\,\chi^{\sigma_2}(Q)\,\tilde{f}_{\tau_1\tau_2}(\sigma_1,\sigma_2),\quad Q \in H,$$

i.e. $\tilde{U}^{\rho_1,\rho_2}_{G_1,G_2}|H$ decomposes into a direct integral of multiples of the irreducible unitary representations of H:

$$(3.1.42)\quad \chi^{\sigma_1}(Q)\,\chi^{\sigma_2}(Q) = \chi^{\sigma_1+\sigma_2}(Q),$$
$$\sigma_1 + \sigma_2 := \begin{cases} (\varkappa_1+\varkappa_2-2\varkappa_1\varkappa_2, \mu_1+\mu_2+\varkappa_1\varkappa_2) & \text{for } H = H_1,\ \sigma_i=(\varkappa_i,\mu_i); \\ (\varkappa_1+\varkappa_2-2\varkappa_1\varkappa_2, \lambda_1+\lambda_2) & \text{for } H = H_2,\ \sigma_i = (\varkappa_i,\lambda_i). \end{cases}$$

With the aid of the substitution

$$(3.1.43)\quad \sigma = \sigma_1 + \sigma_2,\quad \bar{\sigma} = \sigma_1,$$

we may define by

$$(3.1.44)\quad \tilde{f} \longrightarrow f'' = B\tilde{f} : f''_{\tau_1\tau_2}(\sigma,\bar{\sigma}) = \tilde{f}_{\tau_1\tau_2}(\sigma_1,\sigma_2)$$

a unitary map B from $\tilde{\mathfrak{h}}^{\rho_1,\rho_2}_{G_1,G_2,H}$ onto the Hilbert space $\mathfrak{h}''^{\rho_1,\rho_2}_{G_1,G_2,H}$ with scalar product

$$(3.1.45)\quad \langle f''|g''\rangle''^{\rho_1,\rho_2} := \int_{\hat{H}^{\rho_1,\rho_2}} d\hat{\nu}(\sigma) \int_{\hat{H}^{\rho_1,\rho_2}_{\sigma}} d\hat{\nu}(\bar{\sigma}) \sum_{\tau_1=1}^{n(\rho_1)} \sum_{\tau_2=1}^{n(\rho_2)} f''_{\tau_1\tau_2}(\sigma,\bar{\sigma})^* g''_{\tau_1\tau_2}(\sigma,\bar{\sigma}) = \langle \tilde{f}|\tilde{g}\rangle^{\sim\rho_1,\rho_2}.$$

Here the sets $\hat{H}^{\rho_1,\rho_2}$ and $\hat{H}^{\rho_1,\rho_2}_{\sigma}$ are the images of $\hat{H}_{\rho_1} \times \hat{H}_{\rho_2}$ under the substitution(3.1.43). We remark that the Jacobian of the substitution (3.1.43) in any case has the value 1. For the representation

(3.1.46) $$U''^{\rho_1,\rho_2}_{G_1,G_2} := B\, \tilde{U}^{\rho_1,\rho_2}_{G_1,G_2}\, B^{-1}$$

according to (3.1.41) and (3.1.44) holds

(3.1.47)
$$U''^{\rho_1,\rho_2}_{G_1,G_2}(Q) f''_{\tau_1\tau_2}(\sigma,\bar{\sigma}) = \chi^{\sigma}(Q) f''_{\tau_1\tau_2}(\sigma,\bar{\sigma}) \; ,$$
$$U''^{\rho_1,\rho_2}_{G_1,G_2}\Big| H = \oplus\int_{\hat{H}^{\rho_1,\rho_2}} d\hat{\nu}(\sigma) \left[\oplus\int_{\hat{H}^{\rho_1,\rho_2}_{\sigma}} d\hat{\nu}(\bar{\sigma}) \left(\oplus\sum_{\tau_1=1}^{n(\rho_1)} \oplus\sum_{\tau_2=1}^{n(\rho_2)} \chi^{\sigma}\right)\right] \; ,$$

i.e. $U''^{\rho_1,\rho_2}_{G_1,G_2}\Big| H$ decomposes into a direct integral of multiples of irreducible unitary representations of H. The multiplicity of χ^{σ} obviously is given by $n(\rho_1,\rho_2,\sigma)n(\rho_1)n(\rho_2)$ where $n(\rho_1,\rho_2,\sigma)$ is the dimension of the Hilbert space of $\hat{\nu}$-square-integrable, complex valued functions on $\hat{H}^{\rho_1,\rho_2}_{\sigma}$. We now use the composed map $B\circ(\ddot{A}_1 \otimes \ddot{A}_2)$: $\mathfrak{h}^{\rho_1}_{G_1} \otimes \mathfrak{h}^{\rho_2}_{G_2} \longrightarrow \mathfrak{h}''^{\rho_1,\rho_2}_{G_1,G_2,H}$ to map the Hilbert space $\mathfrak{h}'^{1,2}$ from (3.1.33) unitarily onto the Hilbert space

(3.1.48) $$\mathfrak{h}''^{1,2} := \oplus\int_{\Omega_k} \sqrt{d\rho(\mathring{p})} \left[\oplus\int_{\Omega(\mathring{p})} \sqrt{d\omega_{\mathring{p}}(p)} \left(\oplus\int_{\Sigma(p)} \sqrt{dR_p(q)}\, \mathfrak{h}''^{\rho_1,\rho_2}_{G_1,G_2,H}\right)\right] .$$

It consists of the functions $\psi''_{\sigma\bar{\sigma}\tau_1\tau_2}(p,q)$ with scalar product

(3.1.49)
$$\langle\psi''|\varphi''\rangle''^{1,2} := \int_{\Omega_k} d\rho(\mathring{p}) \int_{\Omega(\mathring{p})} d\omega_{\mathring{p}}(p) \int_{\Sigma(p)} dR_p(q) \int_{\hat{H}^{\rho_1,\rho_2}} d\hat{\nu}(\sigma) \int_{\hat{H}^{\rho_1,\rho_2}_{\sigma}} d\hat{\nu}(\bar{\sigma}) \times$$
$$\times \sum_{\tau_1=1}^{n(\rho_1)} \sum_{\tau_2=1}^{n(\rho_2)} \psi''_{\sigma\bar{\sigma}\tau_1\tau_2}(p,q)^* \, \varphi''_{\sigma\bar{\sigma}\tau_1\tau_2}(p,q) = \langle\psi|\varphi\rangle^{1,2} \; .$$

If we carry the product representation $U^{1,2}$ over to $\mathfrak{h}''^{1,2}$ we get according to (3.1.35) and (3.1.47)

(3.1.50)
$$(U''^{1,2}(A,a)\psi'')_{\sigma\bar{\sigma}\tau_1\tau_2}(p,q) := (U^{1,2}(A,a)\psi)''_{\sigma\bar{\sigma}\tau_1\tau_2}(p,q) =$$
$$= e^{ip\cdot a}\chi^{\sigma}(Q(p,q;A))\, \psi''_{\sigma\bar{\sigma}\tau_1\tau_2}(\Lambda(A)^{-1}p, \Lambda(A)^{-1}q) \; ,$$

i.e. $U''^{1,2}$ has the form of a direct integral of representations of $\tilde{P}$ which are induced by irreducible unitary representations of $\check{G}(\mathring{p},\mathring{q})$.

The last step is done with the aid of the Fourier transformation (2.7.7). Since ψ'' according to (3.1.49) for almost all $\bar{\sigma}$ and p is an element of $\oplus\int_{\hat{H}^{\rho_1,\rho_2}}\sqrt{d\hat{\nu}(\sigma)}\,\mathcal{L}^2(\Sigma(p))$, and since $\hat{H}^{\rho_1,\rho_2}\subset\hat{H}$, and $\Sigma(p)$ is homeomorphic with $G(\mathring{p})/H$ we may define by

$$\hat{F}:\mathfrak{h}''^{1,2}\longrightarrow\hat{\mathfrak{h}}'^{1,2},\ \hat{\psi}'=\hat{F}\psi'' ,$$

$$\text{(3.1.51)}\quad \hat{\psi}'^{,\rho\sigma'\tau'}_{\sigma\bar{\sigma}\tau_1\tau_2\tau}(p)=\int_{\Sigma(p)}dR_p(q)\,U^{\rho}_{G(\mathring{p})}(R(\Lambda(p)^{-1}q))_{\tau'\sigma',\tau\sigma}\psi''_{\sigma\bar{\sigma}\tau_1\tau_2}(p,q)\ ,$$

$$\psi''_{\sigma\bar{\sigma}\tau_1\tau_2}(p,q)=\int_{\hat{G}(\mathring{p})_\sigma}d\hat{\mu}(\rho)\int_{\hat{H}_\rho}d\hat{\nu}(\sigma')\sum_{\tau',\tau}\hat{\psi}'^{,\rho\sigma'\tau'}_{\sigma\bar{\sigma}\tau_1\tau_2\tau}(p)\,U^{\rho}_{G(\mathring{p})}(R(\Lambda(p)^{-1}q))^{*}_{\tau'\sigma',\tau\sigma},$$

a unitary map from $\mathfrak{h}''^{1,2}$ onto the Hilbert space

$$\hat{\mathfrak{h}}'^{1,2}=\oplus\int_{\Omega_k}\sqrt{d\rho(\mathring{p})}\ \oplus\int_{G(\mathring{p})^{\rho_1,\rho_2}}\sqrt{d\hat{\mu}(\rho)}\ \hat{\mathfrak{h}}'^{1,2}_{\mathring{p},\rho}\ ,$$

$$\text{(3.1.52)}\quad \begin{aligned}
\hat{\mathfrak{h}}'^{1,2}_{\mathring{p},\rho}: &= \oplus\int_{\hat{H}_\rho^{\rho_1,\rho_2}}\sqrt{d\hat{\nu}(\sigma)}\ \oplus\int_{\hat{H}_\sigma^{\rho_1,\rho_2}}\sqrt{d\hat{\nu}(\bar{\sigma})}\ \oplus\sum_{\tau_1,\tau_2,\tau}\tilde{\mathfrak{h}}^{\mathring{p},\rho}=\mathfrak{h}_\rho^{\rho_1,\rho_2}\otimes\tilde{\mathfrak{h}}^{\mathring{p},\rho}\ ,\\
\tilde{\mathfrak{h}}^{\mathring{p},\rho}: &= \oplus\int_{\Omega(\mathring{p})}\sqrt{d\omega_{\mathring{p}}(p)}\ \oplus\int_{\hat{H}_\rho}\sqrt{d\hat{\nu}(\sigma')}\ \oplus\sum_{\tau'}\mathbb{C}\ ,\\
\mathfrak{h}_\rho^{\rho_1,\rho_2} &= \oplus\int_{\hat{H}_\rho^{\rho_1,\rho_2}}\sqrt{d\hat{\nu}(\sigma)}\ \oplus\int_{\hat{H}_\sigma^{\rho_1,\rho_2}}\sqrt{d\hat{\nu}(\bar{\sigma})}\ \oplus\sum_{\tau_1,\tau_2,\tau}\mathbb{C}\ .
\end{aligned}$$

Here $\hat{G}(\mathring{p})^{\rho_1,\rho_2}$ denotes the set of the $\rho\in\hat{G}(\mathring{p})$ for which $U^{\rho}_{G(\mathring{p})}|H$ contains a χ^σ with $\sigma\in\hat{H}^{\rho_1,\rho_2}$. $\hat{H}_\rho^{\rho_1,\rho_2}$ denotes the set of the $\sigma\in\hat{H}^{\rho_1,\rho_2}$ for which χ^σ occurs in $U^{\rho}_{G(\mathring{p})}|H$. These terms appear in the exchange of the order of integrations:

$$\text{(3.1.53)}\quad \int_{\hat{H}^{\rho_1,\rho_2}}d\hat{\nu}(\sigma)\int_{\hat{G}(\mathring{p})_\sigma}d\hat{\mu}(\rho)=\int_{\hat{G}(\mathring{p})^{\rho_1,\rho_2}}d\hat{\mu}(\rho)\int_{\hat{H}_\rho^{\rho_1,\rho_2}}d\hat{\nu}(\sigma)\ .$$

$\hat{\mathfrak{h}}'^{1,2}$ has the scalar product

$$\langle\hat{\psi}'|\hat{\varphi}'\rangle^{\wedge\prime\,1,2}=\int_{\Omega_k}d\rho(\mathring{p})\int_{\hat{G}(\mathring{p})^{\rho_1,\rho_2}}d\hat{\mu}(\rho)\int_{\hat{H}_\rho^{\rho_1,\rho_2}}d\hat{\nu}(\sigma)\int_{\hat{H}_\sigma^{\rho_1,\rho_2}}d\hat{\nu}(\bar{\sigma})\sum_{\tau_1,\tau_2,\tau}\times$$

$$\text{(3.1.54)}\quad \times\langle\hat{\psi}'^{\rho}_{\sigma\bar{\sigma}\tau_1\tau_2\tau}|\hat{\varphi}'^{\rho}_{\sigma\bar{\sigma}\tau_1\tau_2\tau}\rangle^{\sim\mathring{p},\rho}=\langle\psi|\varphi\rangle^{1,2}\ ,$$

$$\langle\psi'^{\rho}_{\sigma\bar{\sigma}\tau_1\tau_2\tau}|\varphi'^{\rho}_{\sigma\bar{\sigma}\tau_1\tau_2\tau}\rangle^{\sim\mathring{p},\rho}=\int_{\Omega(\mathring{p})}d\omega_{\mathring{p}}(p)\int_{\hat{H}_\rho}d\hat{\nu}(\sigma')\sum_{\tau'}\hat{\psi}'^{,\rho\sigma'\tau'}_{\sigma\bar{\sigma}\tau_1\tau_2\tau}(p)^{*}\,\hat{\varphi}'^{,\rho\sigma'\tau'}_{\sigma\bar{\sigma}\tau_1\tau_2\tau}(p)\ .$$

With the aid of

$$\text{(3.1.55)}\quad U^{\rho}_{G(\mathring{p})}(R(\Lambda(p)^{-1}q)Q(p,q;A))_{\tau'\sigma',\tau\sigma}=U^{\rho}_{G(\mathring{p})}(R(\Lambda(p)^{-1}q))_{\tau'\sigma',\tau\sigma}\times$$
$$\times\chi^{\sigma}(Q(p,q;A)),$$

and the relation

$$(3.1.56) \qquad R(\Lambda(p)^{-1}q)Q(p,q;A) = R(p;A)R(\Lambda(\Lambda^{-1}p)^{-1}\Lambda^{-1}q) \;, \quad \Lambda = \Lambda(A) \;,$$

which follows from (3.1.26), (3.1.21) and (3.1.15), and with the integral formula

$$(3.1.57) \qquad \int_{\Sigma(p)} dR_p(q) = \int_{\Sigma(\Lambda^{-1}p)} dR_{\Lambda^{-1}p}(\Lambda^{-1}q)$$

following from (3.1.11) we can write for the product representation of $\tilde{P}$ after carriing it over to $\hat{\mathfrak{h}}'^{1,2}$

$$(3.1.58) \qquad \begin{aligned} (\hat{U}^{1,2}(A,a)\hat{\psi}')^{\rho\sigma'\tau'}_{\sigma\bar{\sigma}\tau_1\tau_2\tau}(p) :&= (U^{1,2}(A,a)\psi)^{\wedge\prime\rho\sigma'\tau'}_{\sigma\bar{\sigma}\tau_1\tau_2\tau}(p) = \\ &= e^{ip\cdot a}\int_{\hat{H}_\rho} d\hat{\nu}(\sigma'')\sum_{\tau''} U^{\rho}_{G(\mathring{p})}(R(p;A))_{\tau'\sigma',\tau''\sigma''}\hat{\psi}'^{\rho\sigma''\tau''}_{\sigma\bar{\sigma}\tau_1\tau_2\tau}(\Lambda^{-1}p) \;, \end{aligned}$$

i.e. on each Hilbert space $\tilde{\mathfrak{h}}^{\mathring{p},\rho}$ which is contained in the direct integral decomposition of $\hat{\mathfrak{h}}'^{1,2}$ according to (3.1.52) the representation $\hat{U}^{1,2}$ takes the form of an irreducible unitary representation of $\tilde{P}$ of the equivalence class $(\mathring{p},\rho)$. On $\hat{\mathfrak{h}}'^{1,2}_{\mathring{p},\rho}$ then is realized a multiple of this representation with the multiplicity given by the dimension of $\mathfrak{h}^{\rho_1,\rho_2}_{\rho}$. The whole representation space $\hat{\mathfrak{h}}'^{1,2}$ is the direct integral of the Hilbert spaces $\hat{\mathfrak{h}}'^{1,2}_{\mathring{p},\rho}$ over the set of equivalence classes of irreducible unitary representations of $\tilde{P}$ given by Ω_k and $\hat{G}(\mathring{p})^{\rho_1,\rho_2}$.

We give still a somewhat modified form to this solution of the reduction problem for the product representations of $\tilde{P}$ with non vanishing momenta. In (1.1.12) we have defined the Hilbert space $\mathfrak{h}^{\mathring{p},\rho}$ on which the irreducible unitary representation $U^{\mathring{p},\rho}$ of $\tilde{P}$ is realized. According to (3.1.52) and (3.1.36) $\tilde{\mathfrak{h}}^{\mathring{p},\rho}$ is the Hilbert space $\ddot{A}\,\mathfrak{h}^{\mathring{p},\rho}$, equivalent to $\mathfrak{h}^{\mathring{p},\rho}$, on which the restricted representation $U^{\rho}_{G(\mathring{p})}|H$ decomposes. With the aid of the unitary map defined by

$$(3.1.59) \qquad \hat{\psi}'^{\rho}_{\sigma\bar{\sigma}\tau_1\tau_2\tau}(p) \longrightarrow \hat{\psi}^{\rho}_{\sigma\bar{\sigma}\tau_1\tau_2\tau}(p) = \ddot{A}^{-1}\,\hat{\psi}'^{\rho}_{\sigma\bar{\sigma}\tau_1\tau_2\tau}(p)$$

from $\hat{\mathfrak{h}}'^{1,2}$ onto the Hilbert space

$$(3.1.60) \qquad \mathfrak{h}^{1,2} = \oplus\int_{\Omega_k}\sqrt{d\rho(\mathring{p})}\;\oplus\int_{\hat{G}(\mathring{p})^{\rho_1\rho_2}}\sqrt{d\hat{\mu}(\rho)}\;\hat{\mathfrak{h}}^{1,2}_{\mathring{p},\rho} \;, \quad \hat{\mathfrak{h}}^{1,2}_{\mathring{p},\rho} = \mathfrak{h}^{\rho_1,\rho_2}_{\rho}\otimes\mathfrak{h}^{\mathring{p},\rho}$$

we can transform $\hat{U}^{1,2}$ into the representation

$$(\hat{U}^{1,2}(A,a)\hat{\psi})^{\rho}_{\sigma\bar{\sigma}\tau_1\tau_2\tau}(p) := (U^{1,2}(A,a)\psi)^{\wedge\rho}_{\sigma\bar{\sigma}\tau_1\tau_2\tau}(p) =$$

(3.1.61)
$$= e^{ip\cdot a}\, U^{\rho}_{G(\mathring{p})}(R(p;A))\hat{\psi}^{\rho}_{\sigma\bar{\sigma}\tau_1\tau_2\tau}(\Lambda(A)^{-1}p) \ ,$$

$$\hat{U}^{1,2} = \oplus\int_{\Omega_k} d\rho(\mathring{p}) \oplus\int_{\hat{G}(\mathring{p})^{\rho_1,\rho_2}} d\hat{\mu}(\rho) \left[\mathbb{1}_{\mathfrak{h}^{\rho_1,\rho_2}_{\rho}} \otimes U^{\mathring{p},\rho}\right]$$

on $\hat{\mathfrak{h}}^{1,2}$ which may be considered as the final form of the solution of the reduction problem for the product representation $U^{\mathring{p}_1,\rho_1} \otimes U^{\mathring{p}_2,\rho_2}$ of $\tilde{P}$ with non vanishing momenta p_1 and p_2. The domains Ω_k are listed in Table 3.1. The domains $\hat{G}(\mathring{p})^{\rho_1,\rho_2}$ as well as the multiplicity of the representation $U^{\mathring{p},\rho}$ in $\hat{U}^{1,2}$ which is given by the dimension of $\mathfrak{h}^{\rho_1,\rho_2}_{\rho}$, to gether with the Clebsch-Gordan coefficients will be presented in Section 3.2.

3.2 Generalized Clebsch-Gordan Coefficients for the Decomposition of $U^{\mathring{p}_1,\rho_1} \otimes U^{\mathring{p}_2,\rho_2}$, $\mathring{p}_1 \neq 0 \neq \mathring{p}_2$

The unitary transformations (3.1.32), (3.1.38), (3.1.44) and (3.1.51) which lead to the reduction of the product representation $U^{1,2} := U^{\mathring{p}_1,\rho_1} \otimes U^{\mathring{p}_2,\rho_2}$ may be combined to the transformation

$$\hat{\psi}^{\prime\rho\sigma'\tau'}_{\sigma\bar{\sigma}\tau_1'\tau_2'\tau}(p) = \int_{\Sigma(p)} dR_p(q) \int_{\hat{H}_{\rho_1}} d\hat{\nu}(\sigma_1) \int_{\hat{H}_{\rho_2}} d\hat{\nu}(\sigma_2) \sum_{\tau_1\tau_2} \left\langle {\mathring{p}\rho \atop p\tau'\sigma'} ; {\sigma\bar{\sigma} \atop \tau\tau_1'\tau_2'} \middle| {\mathring{p}_1\rho_1 \atop p_1\tau_1\sigma_1} ; {\mathring{p}_2\rho_2 \atop p_2\tau_2\sigma_2} \right\rangle \times$$

(3.2.1)
$$\times\, (\ddot{A}_{1,2}\psi)_{\tau_1\sigma_1,\tau_2\sigma_2}(p_1,p_2) \ ,$$

$$\ddot{A}_{1,2} := \oplus\int_{\Omega(\mathring{p}_1)\times\Omega(\mathring{p}_2)} d\omega_{\mathring{p}_1}(p_1)\, d\omega_{\mathring{p}_2}(p_2)\, (\ddot{A}_1 \otimes \ddot{A}_2)$$

with the generalized Clebsch-Gordan coefficients

$$\left\langle {\mathring{p}\rho \atop p\tau'\sigma'} ; {\sigma\bar{\sigma} \atop \tau\tau_1'\tau_2'} \middle| {\mathring{p}_1\rho_1 \atop p_1\tau_1\sigma_1} ; {p_2\rho_2 \atop p_2\tau_2\sigma_2} \right\rangle := U^{\rho}_{G(\mathring{p})}(R(\Lambda(p)^{-1}q))_{\tau'\sigma',\tau\sigma} \times$$

(3.2.2)
$$\times\, U^{\rho_1}_{G_1}(R(p_1;A(p,q))^{-1})_{\tau_1'\bar{\sigma},\tau_1\sigma_1} U^{\rho_2}_{G_2}(R(p_2;A(p,q))^{-1})_{\tau_2',\sigma-\bar{\sigma},\tau_2\sigma_2} .$$

$A(p)$ and $R(\Lambda(p)^{-1}q)$ may be taken from (1.1.9), (1.1.8) and (3.1.19), (3.1.18), respectively, $A(p,q)$ from (3.1.21) and $R(p_i;A(p,q))$ from (3.1.28) and (3.1.23); p and q as functions of p_1 and p_2 are given by (3.1.1) and (3.1.5). According to (3.1.52) and (3.1.61) the set of irreducible unitary representations of $\tilde{P}$ occurring in $U^{1,2}$ is given by the integration domains Ω_k and $\hat{G}(\mathring{p})^{\rho_1,\rho_2}$, the multiplicity with which

$U^{\mathring{p},\rho}$ occurs in $U^{1,2}$ is given by

$$(3.2.3)\quad d_\rho^{\rho_1,\rho_2} := \dim \mathfrak{h}_\rho^{\rho_1,\rho_2}\ , \quad \mathfrak{h}_\rho^{\rho_1,\rho_2} = \oplus\int_{\hat{H}_\rho^{\rho_1\rho_2}}\sqrt{d\hat{\nu}(\sigma)} \oplus\int_{\hat{H}_\sigma^{\rho_1\rho_2}}\sqrt{d\hat{\nu}(\bar{\sigma})} \oplus \sum_{\tau,\tau_1',\tau_2'} \mathbb{C}\ .$$

The indices σ, $\bar{\sigma}$, τ, τ_1',τ_2' parametrize the degeneracy. In the following for each of the cases I through IX of Table 3.1 we list the irreducible unitary representations contained in $U^{1,2}$ together with their multiplicities. The Clebsch-Gordan coefficients according to (3.2.2) are based on matrix elements of the little group representations. Therefore we cite the formulas of Chapter 2 where the explicit form of these matrix elements is given. The fact that the $\varkappa$-values of commonly appearing equivalence classes ρ and σ are equal is used to write frequently $\sigma = \mu$ or $\sigma = \lambda$ instead of $\sigma = (\varkappa,\mu)$ or $\sigma = (\varkappa,\lambda)$, respectively.

I: $\mathring{p}_1 = m_1 e_{(0)}$, $\mathring{p}_2 = m_2 e_{(0)}$. According to Table 3.1 the domain Ω_I is given by

$$(3.2.4)\qquad \Omega_I = \{me_{(0)}:\ m \geq m_1+m_2\}\ .$$

Therefore $G_1 = G_2 = G(\mathring{p}) = SU(2)$, $G(\mathring{p},\mathring{q}) = H_1$. For $\rho_i = (\varkappa_i,\ell_i)$, $i \in \{1,2\}$ we have $\hat{H}_{1,\rho_i} = \{(\varkappa_i',\mu_i):\ \varkappa_i' = \varkappa_i,\ -\ell_i-\varkappa_i \leq \mu_i \leq \ell_i\}$. The τ-indices may be omitted. The domains $\hat{H}_1^{\rho_1\rho_2}$ and $\hat{H}_{1,\sigma}^{\rho_1\rho_2}$ defined by the substitution (3.1.43) then have the form

$$\hat{H}_1^{\varkappa_1\ell_1,\varkappa_2\ell_2} = \{(\varkappa,\mu):\ \varkappa = \varkappa_1+\varkappa_2-2\varkappa_1\varkappa_2,\ -\ell_1-\varkappa_1/2-\ell_2-\varkappa_2/2 \leq \mu+\varkappa/2 \leq$$
$$\leq \ell_1+\varkappa_1/2+\ell_2+\varkappa_2/2\}\ ,$$

(3.2.5)

$$\hat{H}_{1,\varkappa\mu}^{\varkappa_1\ell_1,\varkappa_2\ell_2} = \{(\bar{\varkappa},\bar{\mu}):\ \bar{\varkappa} = \varkappa_1,\ \max(-\ell_1-\varkappa_1/2,\mu+\varkappa/2-\ell_2-\varkappa_2/2) \leq$$
$$\leq \bar{\mu}+\varkappa_1/2 \leq \min(\ell_1+\varkappa_1/2,\mu+\varkappa/2+\ell_2+\varkappa_2/2)\}\ .$$

Since further

$$(3.2.6)\quad \hat{G}(\mathring{p})_{\varkappa,\mu} = \widehat{SU(2)}_{\varkappa,\mu} = \{(\varkappa',\ell):\ \varkappa' = \varkappa\ ,\ \ell+\varkappa/2 \geq |\mu+\varkappa/2|\}\ ,$$

according to (3.1.53) we get

$$\hat{G}(\mathring{p})^{\varkappa_1\ell_1,\varkappa_2\ell_2} = \{(\varkappa,\ell):\ \varkappa = \varkappa_1+\varkappa_2-2\varkappa_1\varkappa_2,\ \ell \geq 0\}\ ,$$

$$(3.2.7)\quad \hat{H}_{1,\varkappa\ell}^{\varkappa_1\ell_1,\varkappa_2\ell_2} = \{(\varkappa',\mu):\ \varkappa' = \varkappa\ ,\ \max(-\ell-\varkappa/2,-\ell_1-\varkappa_1/2-\ell_2-\varkappa_2/2) \leq$$
$$\leq \mu+\varkappa/2 \leq \min(\ell+\varkappa/2,\ \ell_1+\varkappa_1/2+\ell_2+\varkappa_2/2)\}\ .$$

According to (3.1.52) the multiplicity of the representation $U^{\mathring{p},\varkappa,\ell}$ in $U^{1,2}$ then is given by

$$d^{\varkappa_1\ell_1,\varkappa_2\ell_2}_{\varkappa\ell} = \sum_{\mu=\max(-\ell-\varkappa,-\ell_1-\ell_2-\varkappa_1\varkappa_2-\varkappa)}^{\min(\ell,\ell_1+\ell_2+\varkappa_1\varkappa_2)} \Big[\min(\ell_1,\mu+\ell_2+\varkappa_2-\varkappa_1\varkappa_2) - \max(-\ell_1-\varkappa_1,\mu-\ell_2-\varkappa_1\varkappa_2)+1\Big] =$$

(3.2.8)
$$= \begin{cases} (2\ell+\varkappa+1)\min(2\ell_1+\varkappa_1+1,2\ell_2+\varkappa_2+1) \text{ for } \ell+\varkappa/2 \le |\ell_2+\varkappa_2/2-\ell_1-\varkappa_1/2| \\ (2\ell_1+\varkappa_1+1)(2\ell_2+\varkappa_2+1) - \\ -(\ell+\varkappa/2-\ell_1-\varkappa_1/2-\ell_2-\varkappa_2/2)(\ell+\varkappa/2-\ell_1-\varkappa_1/2-\ell_2-\varkappa_2/2-1) \\ \qquad \text{for } |\ell_2+\varkappa_2/2-\ell_1-\varkappa_1/2| \le \ell+\varkappa/2 \le \ell_2+\varkappa_2/2+\ell_1+\varkappa_1/2 , \\ (2\ell_1+\varkappa_1+1)(2\ell_2+\varkappa_2+1) \text{ for } \ell+\varkappa/2 \ge \ell_1+\varkappa_1/2+\ell_2+\varkappa_2/2 . \end{cases}$$

The Clebsch-Gordan coefficients are

(3.2.9)
$$\left\langle \begin{smallmatrix}\mathring{p}\varkappa\ell\\ p\mu'\end{smallmatrix};\mu\bar{\mu}\middle| \begin{smallmatrix}\mathring{p}_1\varkappa_1\ell_1\\ p_1\mu_1\end{smallmatrix};\begin{smallmatrix}\mathring{p}_2\varkappa_2\ell_2\\ p_2\mu_2\end{smallmatrix}\right\rangle = U^{\varkappa,\ell}_{SU(2)}(R(\Lambda(p)^{-1}q))_{\mu'\mu} \times$$
$$\times U^{\varkappa_1,\ell_1}_{SU(2)}(R(p_1;A(p,q))^{-1})_{\bar{\mu}\mu_1} U^{\varkappa_2,\ell_2}_{SU(2)}(R(p_2;A(p,q))^{-1})_{\mu-\bar{\mu},\mu_2}$$

with the SU(2)-matrix elements specified in (2.2.4).

<u>II: $\mathring{p}_1 = m_1 e_{(0)}$, $\mathring{p}_2 = e_{(0)}+e_{(3)}$.</u> The domain Ω_{II} according to Table 3.1 has the form

(3.2.10)
$$\Omega_{II} = \{me_{(0)}: m > m_1\} .$$

Therefore $G_1 = G(\mathring{p}) = SU(2)$, $G_2 = E(2)$, $G(\mathring{p},\mathring{q}) = H_1$. To $\rho_1 = (\varkappa_1,\ell_1)$ belongs $\hat{H}_{1,\rho_1} = \{(\varkappa_1',\mu_1): \varkappa_1' = \varkappa_1, -\ell_1-\varkappa_1 \le \mu_1 \le \ell_1\}$. With respect to the E(2)-representations we distinguish between a) $\rho_2 = (\varkappa_2,\mu_0)$ with $\hat{H}_{1,\rho_2} = \{(\varkappa_2,\mu_0)\}$ for the onedimensional and b) $\rho_2 = (\varkappa_2,\rho)$ with $\hat{H}_{1,\rho_2} = \{(\varkappa_2',\mu_2): \varkappa_2' = \varkappa_2, -\infty < \mu_2 < \infty\}$ for the infinite-dimensional representations. By the substitution (3.1.43) we get the domains

(3.2.11)
$$\text{a) } \hat{H}_1^{\varkappa_1\ell_1,\varkappa_2\mu_0} = \{(\varkappa,\mu): \varkappa = \varkappa_1+\varkappa_2-2\varkappa_1\varkappa_2,\ \mu_0+\varkappa_2/2-\ell_1-\varkappa_1/2 \le \le \mu+\varkappa/2 \le \mu_0+\varkappa_2/2+\ell_1+\varkappa_1/2\} ,$$

$$\hat{H}^{\varkappa_1 \ell_1, \varkappa_2 \mu_0}_{1,\varkappa\mu} = \{(\bar{\varkappa},\bar{\mu}) : \bar{\varkappa} = \varkappa_1, \bar{\mu} = \mu - \mu_0 - \varkappa_1\varkappa_2\} ,$$

$$\text{(3.2.11) b)}\quad \hat{H}^{\varkappa_1 \ell_1, \varkappa_2 \rho}_{1} = \{(\varkappa,\mu) : \varkappa = \varkappa_1 + \varkappa_2 - 2\varkappa_1\varkappa_2,\ -\infty < \mu < \infty\} ,$$

$$\hat{H}^{\varkappa_1 \ell_1, \varkappa_2 \rho}_{1,\varkappa\mu} = \{(\bar{\varkappa},\bar{\mu}) : \bar{\varkappa} = \varkappa_1,\ -\ell_1 - \varkappa_1 \le \bar{\mu} \le \ell_1\} .$$

Since further

$$\text{(3.2.12)}\quad \hat{G}(\mathring{\mathfrak{p}})_{\varkappa,\mu} = \widehat{SU(2)}_{\varkappa,\mu} = \{(\varkappa',\ell) : \varkappa' = \varkappa,\ \ell \ge |\mu + \varkappa/2| - \varkappa/2\},$$

the relation (3.1.53) yields

$$\text{a)}\quad \hat{G}(\mathring{\mathfrak{p}})^{\varkappa_1 \ell_1, \varkappa_2 \mu_0} = \{(\varkappa,\ell) : \varkappa = \varkappa_1 + \varkappa_2 - 2\varkappa_1\varkappa_2,$$

$$\ell \ge \max(0, |\mu_0 + \varkappa_2/2| - \ell_1 - \varkappa_1/2 - \varkappa/2)\} ,$$

$$\text{(3.2.13)}\quad \hat{H}^{\varkappa_1 \ell_1, \varkappa_2 \mu_0}_{1,\varkappa\ell} = \{(\varkappa',\mu) : \varkappa' = \varkappa,\ \max(-\ell - \varkappa/2, \mu_0 + \varkappa_2/2 - \ell_1 - \varkappa_1/2) \le$$

$$\le \mu + \varkappa/2 \le \min(\ell + \varkappa/2, \mu_0 + \varkappa_2/2 + \ell_1 + \varkappa_1/2)\},$$

$$\text{b)}\quad \hat{G}(\mathring{\mathfrak{p}})^{\varkappa_1 \ell_1, \varkappa_2 \rho} = \{(\varkappa,\ell) : \varkappa = \varkappa_1 + \varkappa_2 - 2\varkappa_1\varkappa_2,\ \ell \ge 0\} ,$$

$$\hat{H}^{\varkappa_1 \ell_1, \varkappa_2 \rho}_{1,\varkappa\ell} = \{(\varkappa',\mu) : \varkappa' = \varkappa,\ -\ell - \varkappa \le \mu \le \ell\} .$$

According to (3.1.52) the multiplicities are given by

$$\text{(3.2.14)}\quad \text{a)}\quad d^{\varkappa_1 \ell_1, \varkappa_2 \mu_0}_{\varkappa\ell} = \begin{cases} 2\ell + \varkappa + 1 \text{ for } \varkappa/2 \le \ell + \varkappa/2 \le \ell_1 + \varkappa_1/2 - |\mu_0 + \varkappa_2/2| , \\ \ell + \varkappa/2 + \ell_1 + \varkappa_1/2 - |\mu_0 + \varkappa_2/2| + 1 \text{ for } \ell_1 + \varkappa_1/2 - \\ \quad - |\mu_0 + \varkappa_2/2| \le \ell + \varkappa/2 \le \ell_1 + \varkappa_1/2 + |\mu_0 + \varkappa_2/2| , \\ 2\ell_1 + \varkappa_1 + 1 \text{ for } \ell + \varkappa/2 \ge \ell_1 + \varkappa_1/2 + |\mu_0 + \varkappa_2/2| , \end{cases}$$

$$\text{b)}\quad d^{\varkappa_1 \ell_1, \varkappa_2 \rho}_{\varkappa\ell} = (2\ell_1 + \varkappa_1 + 1)(2\ell + \varkappa + 1) .$$

The Clebsch-Gordan coefficients are

(3.2.15)

$$\text{a)}\ \left\langle {\mathring{p}\varkappa\ell \atop p\mu'};\mu \middle| {\mathring{p}_1\varkappa_1\ell_1 \atop p_1\mu_1};{\mathring{p}_2\varkappa_2\mu_o \atop p_2}\right\rangle = U^{\varkappa,\ell}_{SU(2)}(R(\Lambda(p)^{-1}q))_{\mu'\mu} \times$$

$$\times\, U^{\varkappa_1,\ell_1}_{SU(2)}(R(p_1;A(p,q))^{-1})_{\mu-\mu_o-\varkappa_1\varkappa_2,\mu_1}\, U^{o,\varkappa_2,\mu_o}_{E(2)}(R(p_2;A(p,q))^{-1}),$$

$$\text{b)}\ \left\langle {\mathring{p}\varkappa\ell \atop p\mu'};\mu\bar{\mu} \middle| {\mathring{p}_1\varkappa_1\ell_1 \atop p_1\mu_1};{\mathring{p}_2\varkappa_2\rho \atop p_2\mu_2}\right\rangle = U^{\varkappa,\ell}_{SU(2)}(R(\Lambda(p)^{-1}q))_{\mu'\mu} \times$$

$$\times\, U^{\varkappa_1,\ell_1}_{SU(2)}(R(p_1;A(p,q))^{-1})_{\bar{\mu}\mu_1}\, U^{\rho,\varkappa_2}_{E(2)}(R(p_2;A(p,q))^{-1})_{\mu-\bar{\mu},\mu_2}$$

with the SU(2)-matrix elements specified in (2.2.4) and the E(2)-matrix elements specified in (1.4.13) and (2.4.4),respectively.

III: $\mathring{p}_1 = m_1 e_{(0)}$, $\mathring{p}_2 = n_2 e_{(3)}$. The domain Ω_{III} according to Table 3.1 has the form

$$(3.2.16)\qquad \Omega_{III} = \{me_{(0)}:\ m>0\}\cup\{ne_{(3)}:\ n>0\}\cup\{e_{(0)}+e_{(3)}\}\ .$$

We remind that the sets $\{\pm e_{(0)}\pm e_{(3)}\}$ and $\{0\}\subset\Omega_k$ need not be considered in the following because in the integral decomposition (3.1.60) of the representation space $\hat{\mathfrak{h}}^{1,2}$ they are of measure zero and therefore negligible.

III.1: $\mathring{p}\in\{me_{(0)}:\ m>0\}$. Then $G_1 = G(\mathring{p}) = SU(2)$, $G_2 = SU(1,1)$, $G(\mathring{p},\mathring{q}) = H_1$. For $\rho_1 = (\varkappa_1,\ell_1)$ we have $\hat{H}_{1,\rho_1} = \{(\varkappa_1',\mu_1):\ \varkappa_1' = \varkappa_1,\ -\ell_1-\varkappa_1\leq\mu_1\leq\ell_1\}$. With respect to the SU(1,1)-representations we distinguish between a) $\rho_2 = (\varkappa_2,\ell_2,o)$ with $\hat{H}_{1,\rho_2} = \{(\varkappa_2',\mu_2):\ \varkappa_2' = \varkappa_2,\ -\infty<\mu_2<\infty\}$ for principal and supplementary series and b) $\rho_2 = (\varkappa_2,\ell_2,\pm)$ with $\hat{H}_{1,\rho_2} = \{(\varkappa_2',\mu_2):\ \varkappa_2' = \varkappa_2,\ \pm(\mu_2+\varkappa_2/2)\geq\ell_2+\varkappa_2/2+1\}$ for the discrete series. By the substitution (3.1.43) we get the domains

(3.2.17)

$$\text{a)}\ \hat{H}_1^{\varkappa_1\ell_1,\varkappa_2\ell_2 o} = \{(\varkappa,\mu):\ \varkappa = \varkappa_1+\varkappa_2-2\varkappa_1\varkappa_2,\ -\infty<\mu<\infty\}\ ,$$

$$\hat{H}_{1,\varkappa\mu}^{\varkappa_1\ell_1,\varkappa_2\ell_2 o} = \{(\bar{\varkappa},\bar{\mu}):\ \bar{\varkappa} = \varkappa_1,-\ell_1-\varkappa_1\leq\bar{\mu}\leq\ell_1\},$$

$$\text{b)}\ \hat{H}_1^{\varkappa_1\ell_1,\varkappa_2\ell_2\pm} = \{(\varkappa,\mu):\ \varkappa = \varkappa_1+\varkappa_2-2\varkappa_1\varkappa_2,\ \pm(\mu+\varkappa/2)\geq\ell_2+\varkappa_2/2-\ell_1-\varkappa_1/2+1\}\ ,$$

$$\hat{H}_{1,\varkappa\mu}^{\varkappa_1\ell_1,\varkappa_2\ell_2\pm} = \{(\bar{\varkappa},\bar{\mu}):\ \bar{\varkappa} = \varkappa_1,\ -\ell_1-\varkappa_1/2\leq\pm(\bar{\mu}+\varkappa_1/2)\leq \min(\ell_1+\varkappa_1/2,\pm(\mu+\varkappa/2)-\ell_2-\varkappa_2/2-1)\}\ .$$

Since further

$$(3.2.18)\quad \hat{G}(\mathring{p})_{\varkappa,\mu} = \widehat{SU(2)}_{\varkappa,\mu} = \{(\varkappa',\ell):\ \varkappa' = \varkappa,\ \ell \geq |\mu+\varkappa/2| - \varkappa/2\}\ ,$$

from (3.1.53) we obtain

$$(3.2.19)\quad \begin{aligned} &a)\ \hat{G}(\mathring{p})^{\varkappa_1\ell_1,\varkappa_2\ell_2 0} = \{(\varkappa,\ell):\varkappa=\varkappa_1+\varkappa_2-2\varkappa_1\varkappa_2,\ \ell\geq 0\}\ ,\\ &b)\ \hat{G}(\mathring{p})^{\varkappa_1\ell_1,\varkappa_2\ell_2\pm} = \{(\varkappa,\ell):\varkappa=\varkappa_1+\varkappa_2-2\varkappa_1\varkappa_2,\ \ell\geq\max(0,\ell_2+\varkappa_2/2-\ell_1-\varkappa_1/2+1)\}\ ,\end{aligned}$$

while $\hat{H}^{\rho_1\rho_2}_{1,\rho}$ is given by

$$(3.2.20)\quad \begin{aligned} &a)\ \hat{H}^{\varkappa_1\ell_1,\varkappa_2\ell_2 0}_{1,\varkappa\ell} = \{(\varkappa',\mu):\ \varkappa' = \varkappa,\ -\ell-\varkappa\leq\mu\leq\ell\}\ ,\\ &b)\ \hat{H}^{\varkappa_1\ell_1,\varkappa_2\ell_2\pm}_{1,\varkappa\ell} = \{(\varkappa',\mu):\ \varkappa' = \varkappa,\ \max(-\ell-\varkappa/2,\\ &\qquad \ell_2+\varkappa_2/2-\ell_1-\varkappa_1/2+1)\leq\pm(\mu+\varkappa/2)\leq\ell+\varkappa/2\}\ .\end{aligned}$$

The multiplicities according to (3.1.52) are

$$a)\ d^{\varkappa_1\ell_1,\varkappa_2\ell_2 0}_{\varkappa\ell} = (2\ell_1+\varkappa_1+1)(2\ell+\varkappa+1)\quad ,$$

$$(3.2.21)\quad b)\ d^{\varkappa_1\ell_1,\varkappa_2\ell_2\pm}_{\varkappa\ell} = \begin{cases} (2\ell+\varkappa+1)(\ell_1+\varkappa_1/2-\ell_2-\varkappa_2/2)\\ \quad \text{for } \varkappa/2\leq\ell+\varkappa/2\leq\ell_1+\varkappa_1/2-\ell_2-\varkappa_2/2-1\ ,\\ \frac{1}{2}(\ell+\varkappa/2+\ell_1+\varkappa_1/2-\ell_2-\varkappa_2/2)\times\\ \qquad\times(\ell+\varkappa/2+\ell_1+\varkappa_1/2-\ell_2-\varkappa_2/2+1)\\ \quad \text{for } \ell_1+\varkappa_1/2-\ell_2-\varkappa_2/2-1\leq\ell+\varkappa/2\leq\\ \qquad\leq\ell_1+\varkappa_1/2+\ell_2+\varkappa_2/2+1\ ,\\ (2\ell_1+\varkappa_1+1)(\ell+\varkappa/2-\ell_2-\varkappa_2/2)\\ \quad \text{for } \ell+\varkappa/2\geq\ell_1+\varkappa_1/2+\ell_2+\varkappa_2/2+1\ .\end{cases}$$

The Clebsch-Gordan coefficients in both cases take the form

$$(3.2.22)\quad \begin{aligned} &\left\langle {\mathring{p}\varkappa\ell \atop p\mu'};\mu\bar{\mu}\middle| {\mathring{p}_1\varkappa_1\ell_1 \atop p_1\mu_1};{\mathring{p}_2\varkappa_2\ell_2\eta_2 \atop p_2\mu_2}\right\rangle = U^{\varkappa,\ell}_{SU(2)}(R(\Lambda(p)^{-1}q))_{\mu'\mu}\times\\ &\times U^{\varkappa_1,\ell_1}_{SU(2)}(R(p_1;A(p,q))^{-1})_{\bar{\mu}\mu_1}U^{\varkappa_2,\ell_2,\eta_2}_{SU(1,1)}(R(p_2;A(p,q))^{-1})_{\mu-\bar{\mu},\mu_2}\end{aligned}$$

with the SU(2)-matrix elements from (2.2.4) and the SU(1,1)-matrix elements from (2.3.15).

III.2: $p \in \{ne_{(3)}: n > 0\}$. Here $G_1 = SU(2)$, $G_2 = G(\mathring{p}) = SU(1,1)$, $G(\mathring{p},\mathring{q}) = H_1$. For $\rho_1 = (\varkappa_1, \ell_1)$ we have $\hat{H}_{1,\rho_1} = \{(\varkappa_1', \mu_1): \varkappa_1' = \varkappa_1, -\ell_1-\varkappa_1 \le \mu_1 \le \ell_1\}$. With respect to the SU(1,1)-representations we again distinguish between a) $\rho_2 = (\varkappa_2, \ell_2, o)$ with $\hat{H}_{1,\rho_2} = \{(\varkappa_2', \mu_2): \varkappa_2' = \varkappa_2, -\infty < \mu_2 < \infty\}$ for principal and supplementary series and b) $\rho_2 = (\varkappa_2, \ell_2, \pm)$ with $\hat{H}_{1,\rho_2} = \{(\varkappa_2', \mu_2): \varkappa_2' = \varkappa_2, \pm(\mu_2+\varkappa_2/2) \ge \ell_2+\varkappa_2/2+1\}$ for the discrete series. The substitution (3.1.43) yields

$$
\begin{aligned}
&\text{a)}\ \hat{H}_1^{\varkappa_1\ell_1,\varkappa_2\ell_2 o} = \{(\varkappa,\mu): \varkappa = \varkappa_1+\varkappa_2-2\varkappa_1\varkappa_2,\ -\infty<\mu<\infty\}\ ,\\
&\quad\ \hat{H}_{1,\varkappa\mu}^{\varkappa_1\ell_1,\varkappa_2\ell_2 o} = \{(\bar\varkappa,\bar\mu): \bar\varkappa = \varkappa_1,\ -\ell_1-\varkappa_1 \le \bar\mu \le \ell_1\}\ ,\\
&\text{b)}\ \hat{H}_1^{\varkappa_1\ell_1,\varkappa_2\ell_2\pm} = \{(\varkappa,\mu): \varkappa = \varkappa_1+\varkappa_2-2\varkappa_1\varkappa_2,\\
&\qquad\qquad \pm(\mu+\varkappa/2) \ge \ell_2+\varkappa_2/2-\ell_1-\varkappa_1/2+1\}\ ,\\
&\quad\ \hat{H}_{1,\varkappa\mu}^{\varkappa_1\ell_1,\varkappa_2\ell_2\pm} = \{(\bar\varkappa,\bar\mu): \bar\varkappa = \varkappa_1,\ -\ell_1-\varkappa_1/2 \le \pm(\bar\mu+\varkappa_1/2) \le\\
&\qquad\qquad \le \min(\ell_1+\varkappa_1/2, \pm(\mu+\varkappa/2)-\ell_2-\varkappa_2/2-1)\}\ .
\end{aligned}
\tag{3.2.23}
$$

Since further

$$
\begin{aligned}
\hat{G}(\mathring{p})_{\varkappa,\mu} = \widehat{SU(1,1)}_{\varkappa,\mu} &= \{(\varkappa',\ell,o): \varkappa' = \varkappa,\ \ell = -(1+\varkappa)/2+ip,\ p \ge 0\}\cup\\
&\cup\{(\varkappa',\ell,\eta): \varkappa' = \varkappa,\ 0 \le \ell \le |\mu+\varkappa/2|-\varkappa/2-1,\ \eta = \operatorname{sign}(\mu+\varkappa/2)\},
\end{aligned}
\tag{3.2.24}
$$

from (3.1.53) we get in both cases

$$
\begin{aligned}
\hat{G}(\mathring{p})^{\varkappa_1\ell_1,\varkappa_2\ell_2\eta_2} &= \{(\varkappa,\ell,o): \varkappa = \varkappa_1+\varkappa_2-2\varkappa_1\varkappa_2,\ \ell = -(1+\varkappa)/2+ip,\ p \ge 0\}\cup\\
&\cup\{(\varkappa,\ell,+): \varkappa = \varkappa_1+\varkappa_2-2\varkappa_1\varkappa_2,\ \ell \ge 0\}\cup\{(\varkappa,\ell,-): \varkappa = \varkappa_1+\varkappa_2-2\varkappa_1\varkappa_2,\ \ell \ge 0\}.
\end{aligned}
\tag{3.2.25}
$$

Here the supplementary series is omitted being of Plancherel measure zero in $\widehat{SU(1,1)}$. $\hat{H}_{1,\rho}^{\rho_1\rho_2}$ is given by

$$
\begin{aligned}
&\text{a)}\ \hat{H}_{1,\varkappa\ell\eta}^{\varkappa_1\ell_1,\varkappa_2\ell_2 o} = \begin{cases} \{(\varkappa',\mu): \varkappa' = \varkappa,\ -\infty<\mu<\infty\} & \text{for } \eta = 0\ ,\\ \{(\varkappa',\mu): \varkappa' = \varkappa,\ \eta(\mu+\varkappa/2) \ge \ell+\varkappa/2+1\} & \text{for } \eta = \pm, \end{cases}\\
&\text{b)}\ \hat{H}_{1,\varkappa\ell o}^{\varkappa_1\ell_1,\varkappa_2\ell_2\pm} = \{(\varkappa',\mu): \varkappa' = \varkappa,\ \pm(\mu+\varkappa/2) \ge \ell_2+\varkappa_2/2-\ell_1-\varkappa_1/2+1\},
\end{aligned}
\tag{3.2.26}
$$

$$(3.2.26)\quad \begin{aligned}\hat{H}^{\varkappa_1 \ell_1, \varkappa_2 \ell_2 \pm}_{1,\varkappa\ell\pm} &= \{(\varkappa',\mu): \varkappa' = \varkappa,\ \pm(\mu+\varkappa/2) \geq \\ &\geq \max(\ell_2+\varkappa_2/2-\ell_1-\varkappa_1/2+1, \ell+\varkappa/2+1)\},\\ \hat{H}^{\varkappa_1 \ell_1, \varkappa_2 \ell_2 \pm}_{1,\varkappa\ell\mp} &= \{(\varkappa',\mu): \varkappa' = \varkappa,\ \ell_2+\varkappa_2/2-\ell_1-\varkappa_1/2+1 \leq \\ &\leq \pm(\mu+\varkappa/2) \leq -\ell-\varkappa/2-1\}.\end{aligned}$$

The multiplicities according to (3.1.52) are

$$\text{a)}\ d^{\varkappa_1 \ell_1, \varkappa_2 \ell_2 0}_{\varkappa \ell \eta} = \chi_0 ,$$

$$(3.2.27)\quad \begin{aligned}\text{b)}\ d^{\varkappa_1 \ell_1, \varkappa_2 \ell_2 \pm}_{\varkappa \ell 0} &= \chi_0 ,\\ d^{\varkappa_1 \ell_1, \varkappa_2 \ell_2 \pm}_{\varkappa \ell \pm} &= \chi_0 ,\\ d^{\varkappa_1 \ell_1, \varkappa_2 \ell_2 \pm}_{\varkappa \ell \mp} &= \tfrac{1}{2}(\ell_1+\varkappa_1/2-\ell_2-\varkappa_2/2-\ell-\varkappa/2-1)(\ell_1+\varkappa_1/2-\ell_2-\varkappa_2/2-\\ &\qquad -\ell-\varkappa/2)\\ &\text{for } \varkappa/2 \leq \ell+\varkappa/2 \leq \ell_1+\varkappa_1/2-\ell_2-\varkappa_2/2-2 .\end{aligned}$$

The Clebsch-Gordan coefficients in both cases take the form

$$(3.2.28)\quad \begin{aligned}\left\langle {\mathring{p}\varkappa\ell\eta \atop p\mu'}; \mu\bar{\mu} \middle| {\mathring{p}_1\varkappa_1\ell_1 \atop p_1\mu_1}; {\mathring{p}_2\varkappa_2\ell_2\eta_2 \atop p_2\mu_2}\right\rangle &= U^{\varkappa,\ell,\eta}_{SU(1,1)}(R(\Lambda(p)^{-1}q))_{\mu'\mu} \times \\ &\times U^{\varkappa_1,\ell_1}_{SU(2)}(R(p_1;A(p,q))^{-1})_{\bar{\mu}\mu_1} U^{\varkappa_2,\ell_2,\eta_2}_{SU(1,1)}(R(p_2;A(p,q))^{-1})_{\mu-\bar{\mu},\mu_2}\end{aligned}$$

with the SU(2)-matrix elements from (2.2.4) and the SU(1,1)-matrix elements from (2.3.15).

<u>IV: $\mathring{p}_1 = m_1 e_{(0)}$, $\mathring{p}_2 = -e_{(0)}-e_{(3)}$.</u> According to Table 3.1 to this case belongs the domain

$$(3.2.29)\quad \Omega_{IV} = \{m e_{(0)}: 0 < m < m_1\} \cup \{n e_{(3)}: n > 0\} \cup \{e_{(0)}+e_{(3)}\} .$$

<u>IV.1: $\mathring{p} \in \{m e_{(0)}: 0 < m < m_1\}$.</u> Then $G_1 = G(\mathring{p}) = SU(2)$, $G_2 = E(2)$, $G(\mathring{p},\mathring{q}) = H_1$. This is exactly the situation dealt with under II. Thus we may take the results from there.

<u>IV.2: $\mathring{p} \in \{n e_{(3)}: n > 0\}$.</u> Here $G_1 = SU(2)$, $G_2 = E(2)$, $G(\mathring{p}) = SU(1,1)$, $G(\mathring{p},\mathring{q}) = H_1$. For $\rho_1 = (\varkappa_1, \ell_1)$ we have $\hat{H}_{1,\rho_1} = \{(\varkappa_1', \mu_1): \varkappa_1' = \varkappa_1, -\ell_1-\varkappa_1 \leq \mu_1 \leq \ell_1\}$. With respect to the E(2)-representations we distinguish between a) $\rho_2 = (\varkappa_2, \mu_0)$ with $\hat{H}_{1,\rho_2} = \{(\varkappa_2, \mu_0)\}$ for the one-dimensional

and b) $\rho_2 = (\varkappa_2, \rho)$ with $\hat{H}_{1,\rho_2} = \{(\varkappa_2', \mu_2): \varkappa_2' = \varkappa_2,\ -\infty < \mu_2 < \infty\}$ for the infinite-dimensional representations. From the substitution (3.1.43) we get

$$\text{a) } \hat{H}_1^{\varkappa_1 l_1, \varkappa_2 \mu_0} = \{(\varkappa,\mu): \varkappa = \varkappa_1 + \varkappa_2 - 2\varkappa_1\varkappa_2,\ \mu_0 + \varkappa_2/2 - l_1 - \varkappa_1/2 \leq \leq \mu + \varkappa/2 \leq \mu_0 + \varkappa_2/2 + l_1 + \varkappa_1/2\},$$

$$\hat{H}_{1,\varkappa\mu}^{\varkappa_1 l_1, \varkappa_2 \mu_0} = \{(\varkappa_1, \mu - \mu_0 - \varkappa_1\varkappa_2)\}, \qquad (3.2.30)$$

$$\text{b) } \hat{H}_1^{\varkappa_1 l_1, \varkappa_2 \rho} = \{(\varkappa,\mu): \varkappa = \varkappa_1 + \varkappa_2 - 2\varkappa_1\varkappa_2,\ -\infty < \mu < \infty\},$$

$$\hat{H}_{1,\varkappa\mu}^{\varkappa_1 l_1, \varkappa_2 \rho} = \{(\bar{\varkappa},\bar{\mu}): \bar{\varkappa} = \varkappa_1,\ -l_1 - \varkappa_1 \leq \bar{\mu} \leq l_1\}.$$

Since further

$$\hat{G}(\mathring{p})_{\varkappa,\mu} = \widehat{SU(1,1)}_{\varkappa,\mu} = \{(\varkappa', l, 0): \varkappa' = \varkappa,\ l = -(1+\varkappa)/2 + ip,\ p \geq 0\} \cup \cup \{(\varkappa', l, \eta): \varkappa' = \varkappa,\ 0 \leq l \leq |\mu + \varkappa/2| - \varkappa/2 - 1,\ \eta = \operatorname{sign}(\mu + \varkappa/2)\}, \qquad (3.2.31)$$

we have from (3.1.53)

$$\text{a) } \hat{G}(\mathring{p})^{\varkappa_1 l_1, \varkappa_2 \mu_0} = \{(\varkappa, l, 0): \varkappa = \varkappa_1 + \varkappa_2 - 2\varkappa_1\varkappa_2,\ l = -(1+\varkappa)/2 + ip,\ p \geq 0\} \cup$$
$$\cup \{(\varkappa, l, +): \varkappa = \varkappa_1 + \varkappa_2 - 2\varkappa_1\varkappa_2,\ \varkappa/2 \leq l + \varkappa/2 \leq l_1 + \varkappa_1/2 + \mu_0 + \varkappa_2/2\} \cup$$
$$\cup \{(\varkappa, l, -): \varkappa = \varkappa_1 + \varkappa_2 - 2\varkappa_1\varkappa_2,\ \varkappa/2 \leq l + \varkappa/2 \leq l_1 + \varkappa_1/2 - \mu_0 - \varkappa_2/2\},$$

$$\hat{H}_{1,\varkappa l 0}^{\varkappa_1 l_1, \varkappa_2 \mu_0} = \{(\varkappa', \mu): \varkappa' = \varkappa,\ \mu_0 + \varkappa_2/2 - l_1 - \varkappa_1/2 \leq \mu + \varkappa/2 \leq \leq \mu_0 + \varkappa_2/2 + l_1 + \varkappa_1/2\},$$

$$\hat{H}_{1,\varkappa l \pm}^{\varkappa_1 l_1, \varkappa_2 \mu_0} = \{(\varkappa', \mu): \varkappa' = \varkappa,\ \max(l + \varkappa/2 + 1, \pm(\mu_0 + \varkappa_2/2) - - l_1 - \varkappa_1/2) \leq \pm(\mu + \varkappa/2) \leq \pm(\mu_0 + \varkappa_2/2) + + l_1 + \varkappa_1/2\}, \qquad (3.2.32)$$

$$\text{b) } \hat{G}(\mathring{p})^{\varkappa_1 l_1, \varkappa_2 \rho} = \{(\varkappa, l, 0): \varkappa = \varkappa_1 + \varkappa_2 - 2\varkappa_1\varkappa_2,\ l = -(1+\varkappa)/2 + ip,\ p \geq 0\} \cup$$
$$\cup \{(\varkappa, l, +): \varkappa = \varkappa_1 + \varkappa_2 - 2\varkappa_1\varkappa_2,\ l \geq 0\} \cup \{(\varkappa, l, -): \varkappa = \varkappa_1 + \varkappa_2 - 2\varkappa_1\varkappa_2,\ l \geq 0\},$$

$$\hat{H}_{1,\varkappa l 0}^{\varkappa_1 l_1, \varkappa_2 \rho} = \{(\varkappa', \mu): \varkappa' = \varkappa,\ -\infty < \mu < \infty\},$$

$$\hat{H}_{1,\varkappa l \pm}^{\varkappa_1 l_1, \varkappa_2 \rho} = \{(\varkappa', \mu): \varkappa' = \varkappa,\ \pm(\mu + \varkappa/2) \geq l + \varkappa/2 + 1\}.$$

The multiplicities according to (3.1.52) are

$$\text{a)}\quad d^{\varkappa_1 \ell_1, \varkappa_2 \mu_0}_{\varkappa \ell 0} = 2\ell_1 + \varkappa_1 + 1 ,$$

$$(3.2.33)\qquad d^{\varkappa_1 \ell_1, \varkappa_2 \mu_0}_{\varkappa \ell \pm} = \begin{cases} 2\ell_1 + \varkappa_1 + 1 \text{ for } \varkappa/2 \leq \ell + \varkappa/2 \leq -\ell_1 - \varkappa_1/2 \pm (\mu_0 + \varkappa_2/2) - 1 , \\ \pm(\mu_0 + \varkappa_2/2) + \ell_1 + \varkappa_1/2 - \ell - \varkappa/2 \text{ for } \pm(\mu_0 + \varkappa_2/2) - \\ \quad -\ell_1 - \varkappa_1/2 - 1 \leq \ell + \varkappa/2 \leq \pm(\mu_0 + \varkappa_2/2) + \ell_1 + \varkappa_1/2 - 1 , \end{cases}$$

$$\text{b)}\quad d^{\varkappa_1 \ell_1, \varkappa_2 \rho}_{\varkappa \ell \eta} = \aleph_0 .$$

The Clebsch-Gordan coefficients take the form

$$\text{a)}\left\langle \begin{matrix} \mathring{p}\varkappa\ell\eta \\ p\,\mu' \end{matrix} ;\mu \middle| \begin{matrix} \mathring{p}_1 \varkappa_1 \ell_1 \\ p_1 \mu_1 \end{matrix} ; \begin{matrix} \mathring{p}_2 \varkappa_2 \mu_0 \\ p_2 \end{matrix} \right\rangle = U^{\varkappa,\ell,\eta}_{SU(1,1)}(R(\Lambda(p)^{-1}q))_{\mu'\mu} \times$$

$$\times U^{\varkappa_1,\ell_1}_{SU(2)}(R(p_1;A(p,q))^{-1})_{\mu-\mu_0-\varkappa_1\varkappa_2,\mu_1} U^{0,\varkappa_2,\mu_0}_{E(2)}(R(p_2;A(p,q))^{-1}),$$

(3.2.34)

$$\text{b)}\left\langle \begin{matrix} \mathring{p}\varkappa\ell\eta \\ p\,\mu' \end{matrix} ;\mu\bar{\mu} \middle| \begin{matrix} \mathring{p}_1 \varkappa_1 \ell_1 \\ p_1 \mu_1 \end{matrix} ; \begin{matrix} \mathring{p}_2 \varkappa_2 \rho \\ p_2 \mu_2 \end{matrix} \right\rangle = U^{\varkappa,\ell,\eta}_{SU(1,1)}(R(\Lambda(p)^{-1}q))_{\mu'\mu} \times$$

$$\times U^{\varkappa_1,\ell_1}_{SU(2)}(R(p_1;A(p,q))^{-1})_{\bar{\mu}\mu_1} U^{\rho,\varkappa_2}_{E(2)}(R(p_2;A(p,q))^{-1})_{\mu-\bar{\mu},\mu_2} ,$$

with the SU(1,1)-matrix elements from (2.3.15), the SU(2)-matrix elements from (2.2.4) and the E(2)-matrix elements from (1.4.13) and (2.4.4) in the cases a) and b), respectively.

V: $\mathring{p}_1 = m_1 e_{(0)}, \mathring{p}_2 = -m_2 e_{(0)}$. According to Table 3.1 the domain Ω_V is given by

$$(3.2.35)\quad \Omega_V = \begin{cases} \{m e_{(0)}: 0 < m < m_1 - m_2\} \cup \{n e_{(3)}: n > 0\} \cup \{e_{(0)} + e_{(3)}\} \text{ for } m_1 > m_2 , \\ \{n e_{(3)}: n > 0\} \cup \{0\} \text{ for } m_1 = m_2 . \end{cases}$$

V.1: $\mathring{p} \in \{m e_{(0)}: 0 < m < m_1 - m_2\}$. Here because of $G_1 = G_2 = G(\mathring{p}) = SU(2)$, $G(\mathring{p},\mathring{q}) = H_1$, the results of case I may be taken over.

V.2: $\mathring{p} \in \{n e_{(3)}: n > 0\}$. Then $G_1 = G_2 = SU(2)$, $G(\mathring{p}) = SU(1,1)$, $G(\mathring{p},\mathring{q}) = H_1$. For $\rho_i = (\varkappa_i, \ell_i)$, $i \in \{1,2\}$ we have $\hat{H}_{1,\rho_i} = \{(\varkappa_i', \mu_i): \varkappa_i' = \varkappa_i, -\ell_i - \varkappa_i \leq \mu_i \leq \ell_i\}$. The substitution (3.1.43) yields

$$(3.2.36)\quad \hat{H}_1^{\varkappa_1 \ell_1, \varkappa_2 \ell_2} = \{(\varkappa,\mu): \varkappa = \varkappa_1 + \varkappa_2 - 2\varkappa_1\varkappa_2 , -\ell_1 - \varkappa_1/2 - \ell_2 - \varkappa_2/2 \leq \leq \mu + \varkappa/2 \leq \ell_1 + \varkappa_1/2 + \ell_2 + \varkappa_2/2\} ,$$

$$\hat{H}_{1,\varkappa\mu}^{\varkappa_1\ell_1,\varkappa_2\ell_2} = \{(\bar{\varkappa},\bar{\mu}) : \bar{\varkappa} = \varkappa_1,\ \max(-\ell_1-\varkappa_1/2,\mu+\varkappa/2-\ell_2-\varkappa_2/2) \leq \leq \bar{\mu}+\varkappa_1/2 \leq \min(\ell_1+\varkappa_1/2,\mu+\varkappa/2+\ell_2+\varkappa_2/2)\} . \tag{3.2.36}$$

Now we have

$$\hat{G}(\mathring{p})_{\varkappa,\mu} = \widehat{SU(1,1)}_{\varkappa,\mu} = \{(\varkappa',\ell,0) : \varkappa' = \varkappa,\ \ell = -(1+\varkappa)/2+ip,\ p \geq 0\} \cup \cup\{(\varkappa',\ell,\eta) : \varkappa' = \varkappa,\ 0 \leq \ell \leq |\mu+\varkappa/2| - \varkappa/2-1,\ \eta = \mathrm{sign}(\mu+\varkappa/2)\} , \tag{3.2.37}$$

and therefore according to (3.1.53)

$$\begin{aligned}
\hat{G}(\mathring{p})^{\varkappa_1\ell_1,\varkappa_2\ell_2} &= \{(\varkappa,\ell,0) : \varkappa = \varkappa_1+\varkappa_2-2\varkappa_1\varkappa_2,\ \ell = -(1+\varkappa)/2+ip, p \geq 0\} \cup \\
&\cup\{(\varkappa,\ell,+) : \varkappa = \varkappa_1+\varkappa_2-2\varkappa_1\varkappa_2,\ 0 \leq \ell \leq \ell_1+\varkappa_1/2+\ell_2+\varkappa_2/2-\varkappa/2-1\} \cup \\
&\cup\{(\varkappa,\ell,-) : \varkappa = \varkappa_1+\varkappa_2-2\varkappa_1\varkappa_2,\ 0 \leq \ell \leq \ell_1+\varkappa_1/2+\ell_2+\varkappa_2/2-\varkappa/2-1\} , \\
\hat{H}_{1,\varkappa\ell 0}^{\varkappa_1\ell_1,\varkappa_2\ell_2} &= \{(\varkappa',\mu) : \varkappa' = \varkappa, -\ell_1-\varkappa_1/2-\ell_2-\varkappa_2/2 \leq \mu+\varkappa/2 \leq \ell_1+\varkappa_1/2+ +\ell_2+\varkappa_2/2\} , \\
\hat{H}_{1,\varkappa\ell\pm}^{\varkappa_1\ell_1,\varkappa_2\ell_2} &= \{(\varkappa',\mu) : \varkappa'=\varkappa,\ \ell+\varkappa/2+1 \leq \pm(\mu+\varkappa/2) \leq \ell_1+\varkappa_1/2+\ell_2+\varkappa_2/2\} .
\end{aligned} \tag{3.2.38}$$

For the multiplicities we get according to (3.1.52)

$$d_{\varkappa\ell 0}^{\varkappa_1\ell_1,\varkappa_2\ell_2} = (2\ell_1+\varkappa_1+1)(2\ell_2+\varkappa_2+1),$$

$$d_{\varkappa\ell\pm}^{\varkappa_1\ell_1,\varkappa_2\ell_2} = \begin{cases} (2\ell_0+\varkappa_0+1)(\ell_1+\varkappa_1/2+\ell_2+\varkappa_2/2-\ell-\varkappa/2-\ell_0-\varkappa_0/2), \\ \quad 2\ell_0+\varkappa_0 := \min(2\ell_1+\varkappa_1, 2\ell_2+\varkappa_2) , \\ \quad \text{for } 0 \leq \ell \leq |\ell_1+\varkappa_1/2-\ell_2-\varkappa_2/2| - \varkappa/2 , \\ \frac{1}{2}(\ell_1+\varkappa_1/2+\ell_2+\varkappa_2/2-\ell-\varkappa/2)(\ell_1+\varkappa_1/2+\ell_2+\varkappa_2/2-\ell-\varkappa/2+1) \\ \quad \text{for } |\ell_1+\varkappa_1/2-\ell_2-\varkappa_2/2| \leq \ell+\varkappa/2 \leq \ell_1+\varkappa_1/2+\ell_2+\varkappa_2/2-1. \end{cases} \tag{3.2.39}$$

The Clebsch-Gordan coefficients are

$$\left\langle {\mathring{p}\varkappa\ell\eta \atop p\mu'} ;\mu\bar{\mu} \middle| {\mathring{p}_1\varkappa_1\ell_1 \atop p_1\mu_1} ; {\mathring{p}_2\varkappa_2\ell_2 \atop p_2\mu_2} \right\rangle = U_{SU(1,1)}^{\varkappa,\ell,\eta}(R(\Lambda(p)^{-1}q))_{\mu'\mu} \times \times U_{SU(2)}^{\varkappa_1,\ell_1}(R(p_1;A(p,q))^{-1})_{\bar{\mu}\mu_1} U_{SU(2)}^{\varkappa_2,\ell_2}(R(p_2;A(p,q))^{-1})_{\mu-\bar{\mu},\mu_2} , \tag{3.2.40}$$

with the SU(1,1)-matrix elements from (2.3.15) and the SU(2)-matrix elements from (2.2.4).

VI: $\mathring{p}_1 = e_{(0)}+e_{(3)}$, $\mathring{p}_2 = e_{(0)}+e_{(3)}$. According to Table 3.1 the domain Ω_{VI} is given by

$$\Omega_{VI} = \{me_{(0)}: m>0\} \cup \{e_{(0)}+e_{(3)}\} . \tag{3.2.41}$$

Therefore $G_1 = G_2 = E(2)$, $G(\mathring{p}) = SU(2)$, $G(\mathring{p},\mathring{q}) = H_1$. We distinguish between a) $\rho_i = (\varkappa_i,\mu_i)$ with $\hat{H}_{1,\rho_i} = \{(\varkappa_i,\mu_i)\}$, $i \in \{1,2\}$ for two one-dimensional, b) $\rho_1 = (\varkappa_1,\mu_1)$, $\rho_2 = (\varkappa_2,\rho_2)$ with $\hat{H}_{1,\rho_1} = \{(\varkappa_1,\mu_1)\}$, $\hat{H}_{1,\rho_2} = \{(\varkappa_2',\mu_2): \varkappa_2' = \varkappa_2, -\infty<\mu_2<\infty\}$ for one onedimensional and one infinite-dimensional, and c) $\rho_i = (\varkappa_i,\rho_i)$ with $\hat{H}_{1,\rho_i} = \{(\varkappa_i',\mu_i): \varkappa_i' = \varkappa_i, -\infty<\mu_i<\infty\}$, $i \in \{1,2\}$ for two infinite-dimensional E(2)-representations. From the substitution (3.1.43) follows

$$\begin{aligned}
&\text{a)}\ \hat{H}_1^{\varkappa_1\mu_1,\varkappa_2\mu_2} = \{(\varkappa,\mu): \varkappa=\varkappa_1+\varkappa_2-2\varkappa_1\varkappa_2,\ \mu=\mu_1+\mu_2+\varkappa_1\varkappa_2\} ,\\
&\quad\ \hat{H}_{1,\varkappa\mu}^{\varkappa_1\mu_1,\varkappa_2\mu_2} = \{(\bar{\varkappa},\bar{\mu}): \bar{\varkappa}=\varkappa_1,\ \bar{\mu}=\mu_1\} ,\\
&\text{b)}\ \hat{H}_1^{\varkappa_1\mu_1,\varkappa_2\rho_2} = \{(\varkappa,\mu): \varkappa=\varkappa_1+\varkappa_2-2\varkappa_1\varkappa_2,\ -\infty<\mu<\infty\} ,\\
&\quad\ \hat{H}_{1,\varkappa\mu}^{\varkappa_1\mu_1,\varkappa_2\rho_2} = \{(\bar{\varkappa},\bar{\mu}): \bar{\varkappa}=\varkappa_1,\ \bar{\mu}=\mu_1\} ,\\
&\text{c)}\ \hat{H}_1^{\varkappa_1\rho_1,\varkappa_2\rho_2} = \{(\varkappa,\mu): \varkappa=\varkappa_1+\varkappa_2-2\varkappa_1\varkappa_2,\ -\infty<\mu<\infty\} ,\\
&\quad\ \hat{H}_{1,\varkappa\mu}^{\varkappa_1\rho_1,\varkappa_2\rho_2} = \{(\bar{\varkappa},\bar{\mu}): \bar{\varkappa}=\varkappa_1,\ -\infty<\bar{\mu}<\infty\} .
\end{aligned} \tag{3.2.42}$$

Because obviously

$$\hat{G}(\mathring{p})_{\varkappa,\mu} = \widehat{SU(2)}_{\varkappa,\mu} = \{(\varkappa',\ell): \varkappa' = \varkappa,\ \ell \geq |\mu+\varkappa/2|-\varkappa/2\} , \tag{3.2.43}$$

we have according to (3.1.53)

$$\begin{aligned}
&\text{a)}\ \hat{G}(\mathring{p})^{\varkappa_1\mu_1,\varkappa_2\mu_2} = \{(\varkappa,\ell): \varkappa=\varkappa_1+\varkappa_2-2\varkappa_1\varkappa_2,\ \ell+\varkappa/2 \geq\\
&\qquad\qquad \geq |\mu_1+\varkappa_1/2+\mu_2+\varkappa_2/2|\} ,\\
&\quad\ \hat{H}_{1,\varkappa\ell}^{\varkappa_1\mu_1,\varkappa_2\mu_2} = \{(\varkappa',\mu): \varkappa'=\varkappa,\ \mu+\varkappa/2 = \mu_1+\varkappa_1/2+\mu_2+\varkappa_2/2\} ,\\
&\text{b)}\ \hat{G}(\mathring{p})^{\varkappa_1\mu_1,\varkappa_2\rho_2} = \{(\varkappa,\ell): \varkappa=\varkappa_1+\varkappa_2-2\varkappa_1\varkappa_2,\ \ell \geq 0\} ,
\end{aligned} \tag{3.2.44}$$

$$\hat{H}_{1,\varkappa\ell}^{\varkappa_1\mu_1,\varkappa_2\rho_2} = \{(\varkappa',\mu): \varkappa' = \varkappa, -\ell-\varkappa \leq \mu \leq \ell\},$$

(3.2.44) c) $$\hat{G}(\mathring{p})^{\varkappa_1\rho_1,\varkappa_2\rho_2} = \{(\varkappa,\ell): \varkappa = \varkappa_1+\varkappa_2-2\varkappa_1\varkappa_2, \ \ell \geq 0\},$$

$$\hat{H}_{1,\varkappa\ell}^{\varkappa_1\rho_1,\varkappa_2\rho_2} = \{(\varkappa',\mu): \varkappa' = \varkappa, -\ell-\varkappa \leq \mu \leq \ell\}.$$

The multiplicities according to (3.1.52) are given by

a) $$d_{\varkappa\ell}^{\varkappa_1\mu_1,\varkappa_2\mu_2} = 1,$$

(3.2.45) b) $$d_{\varkappa\ell}^{\varkappa_1\mu_1,\varkappa_2\rho_2} = 2\ell+\varkappa+1,$$

c) $$d_{\varkappa\ell}^{\varkappa_1\rho_1,\varkappa_2\rho_2} = \aleph_0.$$

The Clebsch-Gordan coefficients are

a) $$\left\langle \begin{matrix}\mathring{p}\varkappa\ell \\ p\mu\end{matrix} \middle| \begin{matrix}\mathring{p}_1\varkappa_1\mu_1; \mathring{p}_2\varkappa_2\mu_2 \\ p_1 \qquad\quad p_2\end{matrix}\right\rangle = U_{SU(2)}^{\varkappa,\ell}(R(\Lambda(p)^{-1}q))_{\mu,\mu_1+\mu_2+\varkappa_1\varkappa_2} \times$$
$$\times\, U_{E(2)}^{0,\varkappa_1,\mu_1}(R(p_1;A(p,q))^{-1})\, U_{E(2)}^{0,\varkappa_2,\mu_2}(R(p_2;A(p,q))^{-1}),$$

b) $$\left\langle \begin{matrix}\mathring{p}\varkappa\ell \\ p\mu'\end{matrix};\mu \middle| \begin{matrix}\mathring{p}_1\varkappa_1\mu_1; \mathring{p}_2\varkappa_2\rho_2 \\ p_1 \qquad\quad p_2\mu_2\end{matrix}\right\rangle = U_{SU(2)}^{\varkappa,\ell}(R(\Lambda(p)^{-1}q))_{\mu'\mu} \times$$

(3.2.46)

$$\times\, U_{E(2)}^{0,\varkappa_1,\mu_1}(R(p_1;A(p,q))^{-1})\, U_{E(2)}^{\rho_2,\varkappa_2}(R(p_2;A(p,q))^{-1})_{\mu-\mu_1,\mu_2},$$

c) $$\left\langle \begin{matrix}\mathring{p}\varkappa\ell \\ p\mu'\end{matrix};\mu\bar{\mu} \middle| \begin{matrix}\mathring{p}_1\varkappa_1\rho_1; \mathring{p}_2\varkappa_2\rho_2 \\ p_1\mu_1 \qquad\quad p_2\mu_2\end{matrix}\right\rangle = U_{SU(2)}^{\varkappa,\ell}(R(\Lambda(p)^{-1}q))_{\mu'\mu} \times$$
$$\times\, U_{E(2)}^{\rho_1,\varkappa_1}(R(p_1;A(p,q))^{-1})_{\bar{\mu}\mu_1}\, U_{E(2)}^{\rho_2,\varkappa_2}(R(p_2;A(p,q))^{-1})_{\mu-\bar{\mu},\mu_2},$$

with the SU(2)-matrix elements from (2.2.4) and the E(2)-matrix elements from (1.4.13) and (2.4.4).

<u>VII: $\mathring{p}_1 = e_{(0)}+e_{(3)},\ \mathring{p}_2 = n_2 e_{(3)}$.</u> According to Table 3.1 to this case belongs the domain

(3.2.47) $$\Omega_{VII} = \{m e_{(0)}: m>0\} \cup \{n e_{(3)}: n>0\} \cup \{e_{(0)}+e_{(3)}\}.$$

<u>VII.1: $\mathring{p} \in \{m e_{(0)}: m>0\}$.</u> Then $G_1 = E(2)$, $G_2 = SU(1,1)$, $G(\mathring{p}) = SU(2)$, $G(\mathring{p},\mathring{q}) = H_1$. We distinguish between a) $\rho_1 = (\varkappa_1,\mu_1)$ with $\hat{H}_{1,\rho_1} = \{(\varkappa_1,\mu_1)\}$ for the onedimensional and b) $\rho_1 = (\varkappa_1,\rho_1)$ with $\hat{H}_{1,\rho_1} = \{(\varkappa_1',\mu_1): \varkappa_1' = \varkappa_1, -\infty<\mu_1<\infty\}$ for the infinite-dimensional E(2)-

representations. For $\rho_2 = (\varkappa_2, \ell_2, \eta_2)$ we have $\hat{H}_{1,\varkappa_2\ell_2 0} = \{(\varkappa_2', \mu_2): \varkappa_2' = \varkappa_2,\ -\infty < \mu_2 < \infty\}$, $\hat{H}_{1,\varkappa_2\ell_2\pm} = \{(\varkappa_2', \mu_2): \varkappa_2' = \varkappa_2,\ \pm(\mu_2+\varkappa_2/2) \geq \geq \ell_2+\varkappa_2/2+1\}$ and therefore with the substitution (3.1.43)

$$\text{a)}\ \hat{H}_1^{\varkappa_1\mu_1,\varkappa_2\ell_2 0} = \{(\varkappa,\mu): \varkappa = \varkappa_1+\varkappa_2-2\varkappa_1\varkappa_2,\ -\infty<\mu<\infty\},$$

$$\hat{H}_1^{\varkappa_1\mu_1,\varkappa_2\ell_2\pm} = \{(\varkappa,\mu): \varkappa = \varkappa_1+\varkappa_2-2\varkappa_1\varkappa_2,\ \pm(\mu+\varkappa/2) \geq$$

$$\geq \ell_2+\varkappa_2/2 \pm (\mu_1+\varkappa_1/2)+1\},$$

$$\hat{H}_{1,\varkappa\mu}^{\varkappa_1\mu_1,\varkappa_2\ell_2\eta_2} = \{(\bar{\varkappa},\bar{\mu}): \bar{\varkappa} = \varkappa_1,\ \bar{\mu} = \mu_1\},$$

(3.2.48)

$$\text{b)}\ \hat{H}_1^{\varkappa_1\rho_1,\varkappa_2\ell_2\eta_2} = \{(\varkappa,\mu): \varkappa = \varkappa_1+\varkappa_2-2\varkappa_1\varkappa_2,\ -\infty<\mu<\infty\},$$

$$\hat{H}_{1,\varkappa\mu}^{\varkappa_1\rho_1,\varkappa_2\ell_2 0} = \{(\bar{\varkappa},\bar{\mu}): \bar{\varkappa} = \varkappa_1,\ -\infty<\bar{\mu}<\infty\},$$

$$\hat{H}_{1,\varkappa\mu}^{\varkappa_1\rho_1,\varkappa_2\ell_2\pm} = \{(\bar{\varkappa},\bar{\mu}): \bar{\varkappa} = \varkappa_1,\ -\infty < \pm(\bar{\mu}+\varkappa_1/2) \leq$$

$$\leq \pm(\mu+\varkappa/2)-\ell_2-\varkappa_2/2-1\}.$$

Obviously we have

$$(3.2.49)\qquad \hat{G}(\overset{\circ}{p})_{\varkappa,\mu} = \widehat{SU(2)}_{\varkappa,\mu} = \{(\varkappa',\ell): \varkappa' = \varkappa,\ \ell \geq |\mu+\varkappa/2|-\varkappa/2\}$$

and therefore according to (3.1.53)

$$\text{a)}\ \hat{G}(\overset{\circ}{p})^{\varkappa_1\mu_1,\varkappa_2\ell_2 0} = \{(\varkappa,\ell): \varkappa = \varkappa_1+\varkappa_2-2\varkappa_1\varkappa_2,\ \ell \geq 0\},$$

$$\hat{H}_{1,\varkappa\ell}^{\varkappa_1\mu_1,\varkappa_2\ell_2 0} = \{(\varkappa',\mu): \varkappa' = \varkappa,\ -\ell-\varkappa \leq \mu \leq \ell\},$$

$$\hat{G}(\overset{\circ}{p})^{\varkappa_1\mu_1,\varkappa_2\ell_2\pm} = \{(\varkappa,\ell): \varkappa = \varkappa_1+\varkappa_2-2\varkappa_1\varkappa_2,$$

$$\ell \geq \max(0, \ell_2+\varkappa_2/2 \pm (\mu_1+\varkappa_1/2)-\varkappa/2+1)\},$$

(3.2.50)

$$\hat{H}_{1,\varkappa\ell}^{\varkappa_1\mu_1,\varkappa_2\ell_2\pm} = \{(\varkappa',\mu): \varkappa' = \varkappa,\ \max(-\ell-\varkappa/2, \ell_2+\varkappa_2/2 \pm$$

$$\pm(\mu_1+\varkappa_1/2)+1) \leq \pm(\mu+\varkappa/2) \leq \ell+\varkappa/2\},$$

$$\text{b)}\ \hat{G}(\overset{\circ}{p})^{\varkappa_1\rho_1,\varkappa_2\ell_2\eta_2} = \{(\varkappa,\ell): \varkappa = \varkappa_1+\varkappa_2-2\varkappa_1\varkappa_2,\ \ell \geq 0\},$$

$$\hat{H}_{1,\varkappa\ell}^{\varkappa_1\rho_1,\varkappa_2\ell_2\eta_2} = \{(\varkappa',\mu): \varkappa' = \varkappa,\ -\ell-\varkappa \leq \mu \leq \ell\}.$$

From (3.1.52) we get the multiplicities

$$\text{a)}\ d_{\varkappa\ell}^{\varkappa_1\mu_1,\varkappa_2\ell_2 o} = 2\ell+\varkappa+1\ ,$$

(3.2.51)
$$d_{\varkappa\ell}^{\varkappa_1\mu_1,\varkappa_2\ell_2\pm} = \begin{cases} 2\ell+\varkappa+1 \text{ for } 0\leq\ell\leq -\ell_2-\varkappa_2/2\mp(\mu_1+\varkappa_1/2)-\varkappa/2-1\ , \\ \ell+\varkappa/2-\ell_2-\varkappa_2/2\mp(\mu_1+\varkappa_1/2) \\ \qquad \text{for } \ell+\varkappa/2\geq-\ell_2-\varkappa_2/2\mp(\mu_1+\varkappa_1/2)-1\ , \end{cases}$$

$$\text{b)}\ d_{\varkappa\ell}^{\varkappa_1\rho_1,\varkappa_2\ell_2\eta_2} = \aleph_0\ .$$

The Clebsch-Gordan coefficients are

(3.2.52)
$$\text{a)}\ \left\langle {\mathring{p}\varkappa\ell \atop p\mu'};\mu \middle| {\mathring{p}_1\varkappa_1\mu_1 \atop p_1};{\mathring{p}_2\varkappa_2\ell_2\eta_2 \atop p_2\mu_2}\right\rangle = U_{SU(2)}^{\varkappa,\ell}(R(\Lambda(p)^{-1}q))_{\mu'\mu}\times$$
$$\times U_{E(2)}^{o,\varkappa_1,\mu_1}(R(p_1;A(p,q))^{-1})\ U_{SU(1,1)}^{\varkappa_2,\ell_2,\eta_2}(R(p_2;A(p,q))^{-1})_{\mu-\mu_1,\mu_2},$$
$$\text{b)}\ \left\langle {\mathring{p}\varkappa\ell \atop p\mu'};\mu\bar{\mu} \middle| {\mathring{p}_1\varkappa_1\rho_1 \atop p_1\mu_1};{\mathring{p}_2\varkappa_2\ell_2\eta_2 \atop p_2\mu_2}\right\rangle = U_{SU(2)}^{\varkappa,\ell}(R(\Lambda(p)^{-1}q))_{\mu'\mu}\times$$
$$\times U_{E(2)}^{\rho_1,\varkappa_1}(R(p_1;A(p,q))^{-1})_{\bar{\mu}\mu_1}\ U_{SU(1,1)}^{\varkappa_2,\ell_2,\eta_2}(R(p_2;A(p,q))^{-1})_{\mu-\bar{\mu},\mu_2}\ ,$$

with the SU(2)-matrix elements from (2.2.4), the E(2)-matrix elements from (1.4.13) and (2.4.4) and the SU(1,1)-matrix elements from (2.3.15). <u>VII.2: $\mathring{p}\in\{ne_{(3)}: n>0\}$.</u> Here G_1 = E(2), $G_2 = G(\mathring{p})$ = SU(1,1), $G(\mathring{p},\mathring{q})$ = = H_1. From VII.1 we take over the sets $\hat{H}_1^{\rho_1\rho_2}$ and $\hat{H}_{1,\varkappa\mu}^{\rho_1\rho_2}$ in (3.2.48). However,

(3.2.53)
$$\hat{G}(\mathring{p})_{\varkappa,\mu} = \widehat{SU(1,1)}_{\varkappa,\mu} = \{(\varkappa',\ell,o):\varkappa'=\varkappa,\ \ell=-(1+\varkappa)/2+ip,\ p\geq 0\}\cup$$
$$\cup\{(\varkappa',\ell,\eta):\varkappa'=\varkappa, 0\leq\ell\leq|\mu+\varkappa/2|-\varkappa/2-1, \eta = \operatorname{sign}(\mu+\varkappa/2)\}\ ,$$

so that (3.1.53) yields

(3.2.54)
$$\text{a)}\ \hat{G}(\mathring{p})^{\varkappa_1\mu_1,\varkappa_2\ell_2 o} = \{(\varkappa,\ell,o):\varkappa=\varkappa_1+\varkappa_2-2\varkappa_1\varkappa_2,\ \ell=-(1+\varkappa)/2+ip, p\geq 0\}\cup$$
$$\cup\{(\varkappa,\ell,+):\varkappa=\varkappa_1+\varkappa_2-2\varkappa_1\varkappa_2,\ \ell\geq 0\}\cup$$
$$\cup\{(\varkappa,\ell,-):\varkappa=\varkappa_1+\varkappa_2-2\varkappa_1\varkappa_2,\ \ell\geq 0\}\ ,$$

$$\hat{G}(\mathring{p})^{\varkappa_1\mu_1,\varkappa_2\ell_2\pm} = \{(\varkappa,\ell,o):\varkappa=\varkappa_1+\varkappa_2-2\varkappa_1\varkappa_2,\ \ell=-(1+\varkappa)/2+ip, p\geq 0\}\cup$$

$$\cup\{(\varkappa,\ell,\pm):\ \varkappa=\varkappa_1+\varkappa_2-2\varkappa_1\varkappa_2,\ \ell\geq 0\}\cup$$
$$\cup\{(\varkappa,\ell,\mp):\ \varkappa=\varkappa_1+\varkappa_2-2\varkappa_1\varkappa_2,\ 0\leq\ell\leq -\ell_2-\varkappa_2/2\mp(\mu_1+\varkappa_1/2)-2\},$$

(3.2.54)

$$\hat{H}_{1,\varkappa\ell o}^{\varkappa_1\mu_1,\varkappa_2\ell_2 o} = \{(\varkappa',\mu):\ \varkappa'=\varkappa,\ -\infty<\mu<\infty\},$$

$$\hat{H}_{1,\varkappa\ell o}^{\varkappa_1\mu_1,\varkappa_2\ell_2\pm} = \{(\varkappa',\mu):\ \varkappa'=\varkappa,\ \pm(\mu+\varkappa/2)\geq\ell_2+\varkappa_2/2\pm \pm(\mu_1+\varkappa_1/2)+1\},$$

$$\hat{H}_{1,\varkappa\ell\pm}^{\varkappa_1\mu_1,\varkappa_2\ell_2 o} = \{(\varkappa',\mu):\ \varkappa'=\varkappa,\ \pm(\mu+\varkappa/2)\geq\ell+\varkappa/2+1\},$$

$$\hat{H}_{1,\varkappa\ell\pm}^{\varkappa_1\mu_1,\varkappa_2\ell_2\pm} = \{(\varkappa',\mu):\ \varkappa'=\varkappa,\ \pm(\mu+\varkappa/2)\geq\max(\ell+\varkappa/2+1,\ \ell_2+\varkappa_2/2+1\pm(\mu_1+\varkappa_1/2))\},$$

$$\hat{H}_{1,\varkappa\ell\mp}^{\varkappa_1\mu_1,\varkappa_2\ell_2\pm} = \{(\varkappa',\mu):\ \varkappa'=\varkappa,\ \ell_2+\varkappa_2/2+1\pm(\mu_1+\varkappa_1/2)\leq \leq\pm(\mu+\varkappa/2)\leq-\ell-\varkappa/2-1\},$$

b) $$\hat{G}(\mathring{p})^{\varkappa_1\rho_1,\varkappa_2\ell_2\eta_2} = \{(\varkappa,\ell,o):\ \varkappa=\varkappa_1+\varkappa_2-2\varkappa_1\varkappa_2,\ \ell=-(1+\varkappa)/2+ip,\ p\geq 0\}\cup$$
$$\cup\{(\varkappa,\ell,+):\ \varkappa=\varkappa_1+\varkappa_2-2\varkappa_1\varkappa_2,\ \ell\geq 0\}\cup$$
$$\cup\{(\varkappa,\ell,-):\ \varkappa=\varkappa_1+\varkappa_2-2\varkappa_1\varkappa_2,\ \ell\geq 0\},$$

$$\hat{H}_{1,\varkappa\ell o}^{\varkappa_1\rho_1,\varkappa_2\ell_2\eta_2} = \{(\varkappa',\mu):\ \varkappa'=\varkappa,\ -\infty<\mu<\infty\},$$

$$\hat{H}_{1,\varkappa\ell\pm}^{\varkappa_1\rho_1,\varkappa_2\ell_2\eta_2} = \{(\varkappa',\mu):\ \varkappa'=\varkappa,\ \pm(\mu+\varkappa/2)\geq\ell+\varkappa/2+1\}.$$

The multiplicities according to (3.1.52) are

(3.2.55)

a) $$d_{\varkappa\ell\eta}^{\varkappa_1\mu_1,\varkappa_2\ell_2\eta_2} = \begin{cases} -\ell-\varkappa/2-\ell_2-\varkappa_2/2\mp(\mu_1+\varkappa_1/2)-1 & \text{for } \eta_2=-\eta=\pm,\\ \aleph_0 & \text{otherwise}, \end{cases}$$

b) $$d_{\varkappa\ell\eta}^{\varkappa_1\rho_1,\varkappa_2\ell_2\eta_2} = \aleph_0 .$$

The Clebsch-Gordan coefficients are

(3.2.56)

a) $$\left\langle {\mathring{p}\varkappa\ell\eta \atop p\mu'};\mu \middle| {\mathring{p}_1\varkappa_1\mu_1 \atop p_1};{\mathring{p}_2\varkappa_2\ell_2\eta_2 \atop p_2\mu_2}\right\rangle = U_{SU(1,1)}^{\varkappa,\ell,\eta}(R(\Lambda(p)^{-1}q))_{\mu'\mu}\ \times$$
$$\times\ U_{E(2)}^{o,\varkappa_1,\mu_1}(R(p_1;A(p,q))^{-1})\ U_{SU(1,1)}^{\varkappa_2,\ell_2,\eta_2}(R(p_2;A(p,q))^{-1})_{\mu-\mu_1,\mu_2},$$

$$\text{(3.2.56)}\quad \text{b)} \left\langle \begin{matrix}\mathring{p}\varkappa\ell\eta\\ p\mu'\end{matrix};\mu\bar{\mu}\middle| \begin{matrix}\mathring{p}_1\varkappa_1\rho_1\\ p_1\mu_1\end{matrix};\begin{matrix}\mathring{p}_2\varkappa_2\ell_2\eta_2\\ p_2\mu_2\end{matrix}\right\rangle = U^{\varkappa,\ell,\eta}_{SU(1,1)}(R(\Lambda(p)^{-1}q))_{\mu'\mu} \times$$

$$\times\, U^{\rho_1,\varkappa_1}_{E(2)}(R(p_1;A(p,q))^{-1})_{\bar{\mu}\mu_1} U^{\varkappa_2,\ell_2,\eta_2}_{SU(1,1)}(R(p_2;A(p,q))^{-1})_{\mu-\bar{\mu},\mu_2} ,$$

with the SU(1,1)-matrix elements from (2.3.15) and the E(2)-matrix elements from (1.4.13) and (2.4.4).

VIII: $\mathring{p}_1 = e_{(0)}+e_{(3)}$, $\mathring{p}_2 = -e_{(0)}-e_{(3)}$. According to Table 3.1 the domain Ω_{VIII} is given by

$$\text{(3.2.57)}\qquad \Omega_{VIII} = \{ne_{(3)}: n > 0\} \cup \{0\} .$$

Then $G_1 = G_2 = E(2)$, $G(\mathring{p}) = SU(1,1)$, $G(\mathring{p},\mathring{q}) = H_1$. We can take over the classification into the cases a), b) and c) of VI with the domains $\hat{H}_1^{\rho_1\rho_2}$ and $\hat{H}^{\rho_1\rho_2}_{1,\varkappa\mu}$ in (3.2.42). With $\hat{G}(\mathring{p})_{\varkappa,\mu} = \widehat{SU(1,1)}_{\varkappa,\mu}$ in (3.2.53) we get from (3.1.53)

$$\text{a)}\ \hat{G}(\mathring{p})^{\varkappa_1\mu_1,\varkappa_2\mu_2} = \{(\varkappa,\ell,o):\varkappa=\varkappa_1+\varkappa_2-2\varkappa_1\varkappa_2,\ell=-(1+\varkappa)/2 +ip,p\geq 0\}\cup$$
$$\cup\{(\varkappa,\ell,\eta):\varkappa=\varkappa_1+\varkappa_2-2\varkappa_1\varkappa_2, 0\leq\ell\leq$$
$$\leq|\mu_1+\varkappa_1/2+\mu_2+\varkappa_2/2|-\varkappa/2-1,$$
$$\eta = \operatorname{sign}(\mu_1+\varkappa_1/2+\mu_2+\varkappa_2/2)\} ,$$

$$\hat{H}^{\varkappa_1\mu_1,\varkappa_2\mu_2}_{1,\varkappa\ell\eta} = \{(\varkappa',\mu):\varkappa'=\varkappa, \mu+\varkappa/2 = \mu_1+\varkappa_1/2+\mu_2+\varkappa_2/2\} ,$$

$$\text{(3.2.58)}\quad \left.\begin{matrix}\text{b)}\\ \text{c)}\end{matrix}\right\} \hat{G}(\mathring{p})^{\varkappa_1\mu_1,\varkappa_2\rho_2} = \hat{G}(\mathring{p})^{\varkappa_1\rho_1,\varkappa_2\rho_2} =$$
$$= \{(\varkappa,\ell,o):\varkappa=\varkappa_1+\varkappa_2-2\varkappa_1\varkappa_2,\ell=-(1+\varkappa)/2+ip,p\geq 0\}\cup$$
$$\cup\{(\varkappa,\ell,+):\varkappa=\varkappa_1+\varkappa_2-2\varkappa_1\varkappa_2, \ell\geq 0\}\cup$$
$$\cup\{(\varkappa,\ell,-):\varkappa=\varkappa_1+\varkappa_2-2\varkappa_1\varkappa_2, \ell\geq 0\} ,$$

$$\hat{H}^{\varkappa_1\mu_1,\varkappa_2\rho_2}_{1,\varkappa\ell o} = \hat{H}^{\varkappa_1\rho_1,\varkappa_2\rho_2}_{1,\varkappa\ell o} = \{(\varkappa',\mu):\varkappa' = \varkappa, -\infty<\mu<\infty\} ,$$

$$\hat{H}^{\varkappa_1\mu_1,\varkappa_2\rho_2}_{1,\varkappa\ell\pm} = \hat{H}^{\varkappa_1\rho_1,\varkappa_2\rho_2}_{1,\varkappa\ell\pm} = \{(\varkappa',\mu): \varkappa' = \varkappa,\pm(\mu+\varkappa/2)\geq\ell+\varkappa/2+1\}.$$

The multiplicities according to (3.1.52) are given by

$$\text{(3.2.59)}\quad \begin{aligned} &\text{a)}\ d^{\varkappa_1\mu_1,\varkappa_2\mu_2}_{\varkappa\ell\eta} = 1 ,\\ &\text{b)}\ d^{\varkappa_1\mu_1,\varkappa_2\rho_2}_{\varkappa\ell\eta} = \aleph_0 ,\\ &\text{c)}\ d^{\varkappa_1\rho_1,\varkappa_2\rho_2}_{\varkappa\ell\eta} = \aleph_0 .\end{aligned}$$

The Clebsch-Gordan coefficients are

$$\text{a)}\ \left\langle {\mathring{p}\varkappa\ell\eta \atop p\mu} \middle| {\mathring{p}_1\varkappa_1\mu_1 \atop p_1};{\mathring{p}_2\varkappa_2\mu_2 \atop p_2}\right\rangle = U^{\varkappa,\ell,\eta}_{SU(1,1)}(R(\Lambda(p)^{-1}q))_{\mu,\mu_1+\mu_2+\varkappa_1\varkappa_2}\times$$
$$\times\, U^{0,\varkappa_1,\mu_1}_{E(2)}(R(p_1;A(p,q))^{-1})\, U^{0,\varkappa_2,\mu_2}_{E(2)}(R(p_2;A(p,q))^{-1}),$$

(3.2.60)
$$\text{b)}\ \left\langle {\mathring{p}\varkappa\ell\eta \atop p\mu'};\mu \middle| {\mathring{p}_1\varkappa_1\mu_1 \atop p_1};{\mathring{p}_2\varkappa_2\rho_2 \atop p_2\mu_2}\right\rangle = U^{\varkappa,\ell,\eta}_{SU(1,1)}(R(\Lambda(p)^{-1}q))_{\mu'\mu}\times$$
$$\times\, U^{0,\varkappa_1,\mu_1}_{E(2)}(R(p_1;A(p,q))^{-1})\, U^{\rho_2,\varkappa_2}_{E(2)}(R(p_2;A(p,q))^{-1})_{\mu-\mu_1,\mu_2},$$

$$\text{c)}\ \left\langle {\mathring{p}\varkappa\ell\eta \atop p\mu'};\mu\bar{\mu} \middle| {\mathring{p}_1\varkappa_1\rho_1 \atop p_1\mu_1};{\mathring{p}_2\varkappa_2\rho_2 \atop p_2\mu_2}\right\rangle = U^{\varkappa,\ell,\eta}_{SU(1,1)}(R(\Lambda(p)^{-1}q))_{\mu'\mu}\times$$
$$\times\, U^{\rho_1,\varkappa_1}_{E(2)}(R(p_1;A(p,q))^{-1})_{\bar{\mu}\mu_1}\, U^{\rho_2,\varkappa_2}_{E(2)}(R(p_2;A(p,q))^{-1})_{\mu-\bar{\mu},\mu_2},$$

with the SU(1,1)-matrix elements from (2.3.15) and the E(2)-matrix elements from (1.4.13) and (2.4.4).

IX: $\mathring{p}_1 = n_1e_{(3)},\ \mathring{p}_2 = n_2e_{(3)}$. According to Table 3.1 the domain Ω_{IX} is given by

(3.2.61)
$$\Omega_{IX} = \{me_{(0)}: m>0\}\cup\{-me_{(0)}: m>0\}\cup\{ne_{(3)}: n>0\}\cup$$
$$\cup\{e_{(0)}+e_{(3)}\}\cup\{-e_{(0)}-e_{(3)}\}\cup \begin{cases}\emptyset & \text{for } n_1\neq n_2,\\ \{0\} & \text{for } n_1=n_2.\end{cases}$$

Then $G_i = SU(1,1)$, $\rho_i = (\varkappa_i,\ell_i,\eta_i)$, $\hat{H}_{1,\varkappa_i\ell_i o} = \{(\varkappa_i',\mu_i): \varkappa_i' = \varkappa_i,\ -\infty<\mu_i<\infty\}$, $\hat{H}_{1,\varkappa_i\ell_i\pm} = \{(\varkappa_i',\mu_i): \varkappa_i' = \varkappa_i,\ \pm(\mu_i+\varkappa_i/2)\geq \ell_i+\varkappa_i/2+1\}$, $i\in\{1,2\}$.

IX.1: $\mathring{p}\in\{me_{(0)}: m>0\}\cup\{-me_{(0)}: m>0\}$. Here $G(\mathring{p}) = SU(2)$, $G(\mathring{p},\mathring{q}) = H_1$. We distinguish between a) $\eta_1 = 0 = \eta_2$, b) $\eta_1 = 0$, $\eta_2 = \pm$, c) $\eta_1 = \pm = \eta_2$ and d) $\eta_1 = \pm = -\eta_2$. The substitution (3.1.43) yields the domains

$$\text{a)}\ \hat{H}_1^{\varkappa_1\ell_1 o,\varkappa_2\ell_2 o} = \{(\varkappa,\mu): \varkappa = \varkappa_1+\varkappa_2-2\varkappa_1\varkappa_2,\ -\infty<\mu<\infty\},$$

$$\hat{H}_{1,}^{\varkappa_1\ell_1 o,\varkappa_2\ell_2 o} = \{(\bar{\varkappa},\bar{\mu}): \bar{\varkappa} = \varkappa_1,\ -\infty<\bar{\mu}<\infty\},$$

(3.2.62)
$$\text{b)}\ \hat{H}_1^{\varkappa_1\ell_1 o,\varkappa_2\ell_2\pm} = \{(\varkappa,\mu): \varkappa = \varkappa_1+\varkappa_2-2\varkappa_1\varkappa_2,\ -\infty<\mu<\infty\},$$

$$\hat{H}_{1,}^{\varkappa_1\ell_1 o,\varkappa_2\ell_2\pm} = \{(\bar{\varkappa},\bar{\mu}): \bar{\varkappa} = \varkappa_1,\ \pm(\bar{\mu}+\varkappa_1/2)\leq\pm(\mu+\varkappa/2)-\ell_2-\varkappa_2/2-1\},$$

$$\text{c)}\ \hat{H}_1^{\varkappa_1\ell_1\pm,\varkappa_2\ell_2\pm} = \{(\varkappa,\mu): \varkappa = \varkappa_1+\varkappa_2-2\varkappa_1\varkappa_2,\ \pm(\mu+\varkappa/2)\geq$$

$$\geq l_1+\varkappa_1/2+l_2+\varkappa_2/2+2\} ,$$

(3.2.62)

$$\hat{H}_{1,\varkappa\mu}^{\varkappa_1 l_1 \pm, \varkappa_2 l_2 \pm} = \{(\bar{\varkappa},\bar{\mu}) : \bar{\varkappa}=\varkappa_1 , l_1+\varkappa_1/2+1 \leq \pm(\bar{\mu}+\varkappa_1/2) \leq \leq \pm(\mu+\varkappa/2)-l_2-\varkappa_2/2-1\} ,$$

$$\text{d)}\ \hat{H}_1^{\varkappa_1 l_1 \pm, \varkappa_2 l_2 \mp} = \{(\varkappa,\mu) : \varkappa=\varkappa_1+\varkappa_2-2\varkappa_1\varkappa_2 , -\infty<\mu<\infty\} ,$$

$$\hat{H}_{1,}^{\varkappa_1 l_1 \pm, \varkappa_2 l_2 \mp} = \{(\bar{\varkappa},\bar{\mu}) : \bar{\varkappa}=\varkappa_1 , \pm(\bar{\mu}+\varkappa_1/2) \geq \geq \max(l_1+\varkappa_1/2+1, \pm(\mu+\varkappa/2)+l_2+\varkappa_2/2+1)\} .$$

Since in any case

$$\text{(3.2.63)} \quad \hat{G}(\mathring{p})_{\varkappa,\mu} = \widehat{SU(2)}_{\varkappa,\mu} = \{(\varkappa',l) : \varkappa' = \varkappa, l \geq |\mu+\varkappa/2|-\varkappa/2\} ,$$

from (3.1.53) follows

(3.2.64)

$$\left.\begin{matrix}\text{a)}\\ \text{b)}\\ \text{d)}\end{matrix}\right\} \hat{G}(\mathring{p})^{\varkappa_1 l_1 \eta_1, \varkappa_2 l_2 \eta_2} = \{(\varkappa,l) : \varkappa=\varkappa_1+\varkappa_2-2\varkappa_1\varkappa_2 , l\geq 0\} ,$$

$$\hat{H}_{1,\varkappa l}^{\varkappa_1 l_1 \eta_1, \varkappa_2 l_2 \eta_2} = \{(\varkappa',\mu) : \varkappa' = \varkappa, -l-\varkappa \leq \mu \leq l\} ,$$

$$\text{c)}\ \hat{G}(\mathring{p})^{\varkappa_1 l_1 \pm, \varkappa_2 l_2 \pm} = \{(\varkappa,l) : \varkappa=\varkappa_1+\varkappa_2-2\varkappa_1\varkappa_2 , l+\varkappa/2 \geq \geq l_1+\varkappa_1/2+l_2+\varkappa_2/2+2\} ,$$

$$\hat{H}_{1,\varkappa l}^{\varkappa_1 l_1 \pm, \varkappa_2 l_2 \pm} = \{(\varkappa',\mu) : \varkappa' = \varkappa, l_1+\varkappa_1/2+l_2+\varkappa_2/2+2 \leq \leq \pm(\mu+\varkappa/2) \leq l+\varkappa/2\} .$$

The multiplicities according to (3.1.52) are given by

(3.2.65)

$$\left.\begin{matrix}\text{a)}\\ \text{b)}\\ \text{d)}\end{matrix}\right\} d_{\varkappa l}^{\varkappa_1 l_1 \eta_1, \varkappa_2 l_2 \eta_2} = \chi_0 ,$$

$$\text{c)}\ d_{\varkappa l}^{\varkappa_1 l_1 \pm, \varkappa_2 l_2 \pm} = \frac{1}{2}(l+\varkappa/2-l_1-\varkappa_1/2-l_2-\varkappa_2/2-1) \times \times (l+\varkappa/2-l_1-\varkappa_1/2-l_2-\varkappa_2/2) .$$

The Clebsch-Gordan coefficients are

(3.2.66)

$$\left\langle {\mathring{p}\varkappa l \atop p\mu'} ;\mu\bar{\mu} \middle| {\mathring{p}_1\varkappa_1 l_1\eta_1 \atop p_1\mu_1} ; {\mathring{p}_2\varkappa_2 l_2\eta_2 \atop p_2\mu_2} \right\rangle = U_{SU(2)}^{\varkappa,l}(R(\Lambda(p)^{-1}q))_{\mu'\mu} \times$$

$$\times U_{SU(1,1)}^{\varkappa_1,l_1,\eta_1}(R(p_1;A(p,q))^{-1})_{\bar{\mu}\mu_1} U_{SU(1,1)}^{\varkappa_2,l_2,\eta_2}(R(p_2;A(p,q))^{-1})_{\mu-\bar{\mu},\mu_2}$$

with the SU(2)-matrix elements from (2.2.4) and the SU(1,1)-matrix elements from (2.3.15).

IX.2: $\mathring{p} \in \{ne_{(3)}: n > 0\}$. Here $G(\mathring{p}) = SU(1,1)$. We divide once more into two subcases.

IX.2_1: $\mathring{q} \in \{me_{(0)}: m > 0\} \cup \{-me_{(0)}: m > 0\}$. Then $G(\mathring{p},\mathring{q}) = H_1$. We take over the classification into the cases a) - d) from IX.1 as well as the sets $\hat{H}_1^{\rho_1\rho_2}$ and $\hat{H}_{1,\varkappa\mu}^{\rho_1\rho_2}$ from (3.2.62). Since in any case

$$\hat{G}(\mathring{p})_{\varkappa,\mu} = \widehat{SU(1,1)}_{\varkappa,\mu} = \{(\varkappa',\ell,o): \varkappa'=\varkappa,\ \ell=-(1+\varkappa)/2+ip,\ p \geq 0\} \cup \tag{3.2.67}$$
$$\cup \{(\varkappa',\ell,\eta): \varkappa'=\varkappa,\ 0 \leq \ell \leq |\mu+\varkappa/2| - \varkappa/2-1,\ \eta = \mathrm{sign}(\mu+\varkappa/2)\},$$

we get from (3.1.53)

$$\left.\begin{matrix} a) \\ b) \\ d) \end{matrix}\right\} \hat{G}(\mathring{p})^{\varkappa_1\ell_1\eta_1,\varkappa_2\ell_2\eta_2} = \{(\varkappa,\ell,o): \varkappa=\varkappa_1+\varkappa_2-2\varkappa_1\varkappa_2,$$
$$\ell = -(1+\varkappa)/2+ip,\ p \geq 0\} \cup$$
$$\cup \{(\varkappa,\ell,+): \varkappa=\varkappa_1+\varkappa_2-2\varkappa_1\varkappa_2,\ \ell \geq 0\} \cup$$
$$\cup \{(\varkappa,\ell,-): \varkappa=\varkappa_1+\varkappa_2-2\varkappa_1\varkappa_2,\ \ell \geq 0\},$$

$$\hat{H}_{1,\varkappa\ell o}^{\varkappa_1\ell_1\eta_1,\varkappa_2\ell_2\eta_2} = \{(\varkappa',\mu): \varkappa'=\varkappa,\ -\infty < \mu < \infty\},$$

$$\hat{H}_{1,\varkappa\ell\pm}^{\varkappa_1\ell_1\eta_1,\varkappa_2\ell_2\eta_2} = \{(\varkappa',\mu): \varkappa'=\varkappa,\ \pm(\mu+\varkappa/2) \geq \ell+\varkappa/2+1\},$$

(3.2.68)

$$c)\ \hat{G}(\mathring{p})^{\varkappa_1\ell_1\pm,\varkappa_2\ell_2\pm} = \{(\varkappa,\ell,o): \varkappa=\varkappa_1+\varkappa_2-2\varkappa_1\varkappa_2,$$
$$\ell = -(1+\varkappa)/2+ip,\ p \geq 0\} \cup$$
$$\cup \{(\varkappa,\ell,\pm): \varkappa=\varkappa_1+\varkappa_2-2\varkappa_1\varkappa_2,\ \ell \geq 0\},$$

$$\hat{H}_{1,\varkappa\ell o}^{\varkappa_1\ell_1\pm,\varkappa_2\ell_2\pm} = \{(\varkappa',\mu): \varkappa'=\varkappa, \pm(\mu+\varkappa/2) \geq \ell_1+\varkappa_1/2+\ell_2+\varkappa_2/2+2\},$$

$$\hat{H}_{1,\varkappa\ell\pm}^{\varkappa_1\ell_1\pm,\varkappa_2\ell_2\pm} = \{(\varkappa',\mu): \varkappa'=\varkappa, \pm(\mu+\varkappa/2) \geq$$
$$\geq \max(\ell+\varkappa/2+1,\ \ell_1+\varkappa_1/2+\ell_2+\varkappa_2/2+2)\}.$$

The multiplicity according to (3.1.52) in any case is

$$d_{\varkappa\ell\eta}^{\varkappa_1\ell_1\eta_1,\varkappa_2\ell_2\eta_2} = \chi_o. \tag{3.2.69}$$

The Clebsch-Gordan coefficients have the form

$$\left\langle {}^{\beta\varkappa\ell\eta}_{p\mu'};\mu\bar{\mu}\middle| {}^{\beta_1\varkappa_1\ell_1\eta_1}_{p_1\mu_1};{}^{\beta_2\varkappa_2\ell_2\eta_2}_{p_2\mu_2}\right\rangle = U^{\varkappa,\ell,\eta}_{SU(1,1)}(R(\Lambda(p)^{-1}q))_{\mu'\mu}\times \tag{3.2.70}$$
$$\times U^{\varkappa_1,\ell_1,\eta_1}_{SU(1,1)}(R(p_1;A(p,q))^{-1})_{\bar{\mu}\mu_1}U^{\varkappa_2,\ell_2,\eta_2}_{SU(1,1)}(R(p_2;A(p,q))^{-1})_{\mu-\bar{\mu},\mu_2},$$

with the SU(1,1)-matrix elements from (2.3.15).

IX.2_2: $\mathring{q}\in\{ne_{(2)}: n>0\}$. Here $G(\mathring{p},\mathring{q}) = H_2$. For $\rho_i = (\varkappa_i,\ell_i,\eta_i)$ we have $\hat{H}_{2,\varkappa_i\ell_i\eta_i} = \{(\varkappa_i',\lambda_i): \varkappa_i' = \varkappa_i, \lambda_i\in\mathbb{R}\}$, $i\in\{1,2\}$. The τ-indices take the values $\pm$. The substitution (3.1.43) yields

$$\begin{aligned} \hat{H}_2^{\varkappa_1\ell_1\eta_1,\varkappa_2\ell_2\eta_2} &= \{(\varkappa,\lambda): \varkappa = \varkappa_1+\varkappa_2-2\varkappa_1\varkappa_2, -\infty<\lambda<\infty\}, \\ \hat{H}_{2,\varkappa\lambda}^{\varkappa_1\ell_1\eta_1,\varkappa_2\ell_2\eta_2} &= \{(\bar{\varkappa},\bar{\lambda}): \bar{\varkappa} = \varkappa_1, -\infty<\bar{\lambda}<\infty\}. \end{aligned} \tag{3.2.71}$$

Obviously

$$\begin{aligned} \hat{G}(\mathring{p})_{\varkappa,\lambda} = \widehat{SU(1,1)}_{\varkappa,\lambda} &= \{(\varkappa',\ell,0): \varkappa'=\varkappa, \ell=-(1+\varkappa)/2+ip, p\geq 0\}\cup \\ &\cup\{(\varkappa',\ell,+): \varkappa'=\varkappa, \ell\geq 0\}\cup\{(\varkappa',\ell,-): \varkappa'=\varkappa, \ell\geq 0\} \end{aligned} \tag{3.2.72}$$

and therefore according to (3.1.53)

$$\begin{aligned} \hat{G}(\mathring{p})^{\varkappa_1\ell_1\eta_1,\varkappa_2\ell_2\eta_2} &= \{(\varkappa,\ell,0): \varkappa=\varkappa_1+\varkappa_2-2\varkappa_1\varkappa_2, \ell=-(1+\varkappa)/2+ip, p\geq 0\}\cup \\ &\cup\{(\varkappa,\ell,+): \varkappa=\varkappa_1+\varkappa_2-2\varkappa_1\varkappa_2, \ell\geq 0\}\cup \\ &\cup\{(\varkappa,\ell,-): \varkappa=\varkappa_1+\varkappa_2-2\varkappa_1\varkappa_2, \ell\geq 0\}, \\ \hat{H}_{2,\varkappa\ell\eta}^{\varkappa_1\ell_1\eta_1,\varkappa_2\ell_2\eta_2} &= \{(\varkappa',\lambda): \varkappa' = \varkappa, -\infty<\lambda<\infty\}. \end{aligned} \tag{3.2.73}$$

The multiplicity of $U^{\varkappa,\ell,\eta}_{SU(1,1)}$ in $U^{1,2}$ according to (3.1.52) is given by the countably infinite dimension of the Hilbert space $\mathfrak{h}^{\varkappa_1\ell_1\eta_1,\varkappa_2\ell_2\eta_2}_{\varkappa\ell\eta}$:

$$d^{\varkappa_1\ell_1\eta_1,\varkappa_2\ell_2\eta_2}_{\varkappa\ell\eta} = \aleph_0 . \tag{3.2.74}$$

The Clebsch-Gordan coefficients are

$$\begin{aligned} \left\langle {}^{\beta\varkappa\ell\eta}_{p\tau'\lambda'};{}^{\lambda\bar{\lambda}}_{\tau\tau_1'\tau_2'}\middle| {}^{\beta_1\varkappa_1\ell_1\eta_1}_{p_1\tau_1\lambda_1};{}^{\beta_2\varkappa_2\ell_2\eta_2}_{p_2\tau_2\lambda_2}\right\rangle &= U^{\varkappa,\ell,\eta}_{SU(1,1)}(R(\Lambda(p)^{-1}q))_{\tau'\lambda',\tau\lambda}\times \\ &\times U^{\varkappa_1,\ell_1,\eta_1}_{SU(1,1)}(R(p_1;A(p,q))^{-1})_{\tau_1'\bar{\lambda},\tau_1\lambda_1}\times \\ &\times U^{\varkappa_2,\ell_2,\eta_2}_{SU(1,1)}(R(p_2;A(p,q))^{-1})_{\tau_2',\lambda-\bar{\lambda},\tau_2\lambda_2}. \end{aligned} \tag{3.2.75}$$

Here the SU(1,1)-matrix elements in the H_2-basis are given by (2.5.74) and (2.5.62) together with the symmetry relations (2.5.66) or (2.5.72).

3.3 Remarks on the Reduction of the Product Representations $U^{0,\rho_1} \otimes U^{\mathring{p},\rho_2}$ of $\tilde{P}$

The product representation $U^{1,2} \equiv U^{0,\rho_1} \otimes U^{p,\rho_2}$ according to (3.1), (3.4) and (1.1.16) has the form

$$(3.3.1)\quad (U^{1,2}(A,a)\psi)(p) = e^{ip\cdot a}\, U^{\rho_1}_{SL(2,\mathbb{C})}(A) \otimes U^{\rho_2}_{G(\mathring{p})}(R(p;A))\psi(\Lambda(A)^{-1}p).$$

In the following we present a construction which leads to the reduction of $U^{1,2}$. However, the reduction problems for unitary representations of the little groups which occur in this process are not solved explicitly, but the literature handling with these problems is cited as far as known to us.

Let be $\mathring{p} \neq 0$. Then by the unitary transformation

$$(3.3.2)\quad \psi \longrightarrow \psi' :\ \psi'(p) = (U^{\rho_1}_{SL(2,\mathbb{C})}(A(p)^{-1}) \otimes \mathbb{1}_{\mathfrak{h}^{\rho_2}_{G(\mathring{p})}})\,\psi(p)$$

$U^{1,2}$ is carried over to the representation $U'^{1,2}$ with

$$(3.3.3)\quad \begin{aligned} (U'^{1,2}(A,a)\psi')(p) &:= (U^{1,2}(A,a)\psi)'(p) = \\ &= e^{ip\cdot a}\, U^{\rho_1}_{SL(2,\mathbb{C})}(R(p;A)) \otimes U^{\rho_2}_{G(\mathring{p})}(R(p;A))\psi'(\Lambda(A)^{-1}p). \end{aligned}$$

This has the form of a representation of $\tilde{P}$ that is induced by the representation $U^{0,\rho_1}_{\tilde{P}}|\check{G}(\mathring{p}) \otimes U^{\mathring{p},\rho_2}_{\check{G}(\mathring{p})}$ of $\check{G}(\mathring{p}) = G(\mathring{p}) \circledS \mathbb{R}^4 \subset \tilde{P}$. The following two steps lead to the solution of the reduction problem for $U^{1,2}$

<u>1.</u> By a unitary transformation $\ddot{A}^{\rho_1}$ the representation space $\mathfrak{h}^{\rho_1}_{SL(2,\mathbb{C})}$ is mapped onto the Hilbert space

$$(3.3.4)\quad \ddot{A}^{\rho_1}\, \mathfrak{h}^{\rho_1}_{SL(2,\mathbb{C})} = \oplus\!\int_{\hat{G}(\mathring{p})} \sqrt{d\tilde{\mu}_{\rho_1}(\rho)}\ \mathfrak{h}^{\rho_1}_{\rho} \otimes \mathfrak{h}^{\rho}_{G(\mathring{p})}\ ,$$

on which the restricted representation $U^{\rho_1}_{SL(2,\mathbb{C})}|G(\mathring{p})$ decomposes into the direct integral

$$(3.3.5)\quad \ddot{A}^{\rho_1}\, U^{\rho_1}_{SL(2,\mathbb{C})}|G(\mathring{p})\ \ddot{A}^{\rho_1\,-1} = \oplus\!\int_{\hat{G}(\mathring{p})} d\tilde{\mu}_{\rho_1}(\rho)\ (\mathbb{1}_{\mathfrak{h}^{\rho_1}_{\rho}} \otimes U^{\rho}_{G(\mathring{p})})$$

of irreducible unitary representations $U^{\rho}_{G(\mathfrak{p})}$ of $G(\mathfrak{p})$ each occurring with the multiplicity given by the dimension of $\mathfrak{h}^{\rho_1}_{\rho}$. The measure $\tilde{\mu}_{\rho_1}$ on the set $\hat{G}(\mathfrak{p})$ of equivalence classes of irreducible unitary representations of $G(\mathfrak{p})$ is unique up to equivalence. Now the Hilbert space

$$(3.3.6)\quad A^{\rho_1}\,\mathfrak{h}^{\rho_1}_{SL(2,\mathbb{C})} \otimes \mathfrak{h}^{\rho_2}_{G(\mathfrak{p})} = \left[\oplus\int_{\hat{G}(\mathfrak{p})} \sqrt{d\tilde{\mu}_{\rho_1}(\rho)}\,(\mathfrak{h}^{\rho_1}_{\rho} \otimes \mathfrak{h}^{\rho}_{G(\mathfrak{p})})\right] \otimes \mathfrak{h}^{\rho_2}_{G(\mathfrak{p})}$$

is mapped unitarily by

$$(3.3.7)\quad \begin{gathered} \psi' = \Big(\oplus\int_{\hat{G}(\mathfrak{p})} \sqrt{d\tilde{\mu}_{\rho_1}(\rho)}\,\psi'^{\rho_1}_{\rho}\Big) \otimes \psi^{\rho_2} \longrightarrow \tilde{\psi} := \oplus\int_{\hat{G}(\mathfrak{p})} \sqrt{d\tilde{\mu}_{\rho_1}(\rho)}\,(\psi'^{\rho_1}_{\rho} \otimes \psi^{\rho_2}), \\ \psi'^{\rho_1}_{\rho} \in \mathfrak{h}^{\rho_1}_{\rho} \otimes \mathfrak{h}^{\rho}_{G(\mathfrak{p})},\quad \psi^{\rho_2} \in \mathfrak{h}^{\rho_2}_{G(\mathfrak{p})}, \end{gathered}$$

onto the Hilbert space

$$(3.3.8)\quad \tilde{\mathfrak{h}}^{1,2} := \oplus\int_{\hat{G}(\mathfrak{p})} \sqrt{d\tilde{\mu}_{\rho_1}(\rho)}\,(\mathfrak{h}^{\rho_1}_{\rho} \otimes \mathfrak{h}^{\rho}_{G(\mathfrak{p})} \otimes \mathfrak{h}^{\rho_2}_{G(\mathfrak{p})})$$

(referring to this cf. DIXMIER [30]). Then the representation $U'^{1,2}$ from (3.3.3) is carried over to

$$(3.3.9)\quad \begin{aligned} (\tilde{U}^{1,2}(A,a)\tilde{\psi})(p) &:= (U^{1,2}(A,a)\psi)^{\sim}(p) = \\ &= e^{ip\cdot a} \oplus\int_{\hat{G}(\mathfrak{p})} d\tilde{\mu}_{\rho_1}(\rho)\,\Big[\mathbb{1}_{\mathfrak{h}^{\rho_1}_{\rho}} \otimes U^{\rho}_{G(\mathfrak{p})}(R(p;A)) \otimes \\ &\qquad \otimes U^{\rho_2}_{G(\mathfrak{p})}(R(p;A))\Big]\tilde{\psi}(\Lambda(A)^{-1}p)\ . \end{aligned}$$

2. By a unitary transformation $A^{\rho\rho_2}$ the product space $\mathfrak{h}^{\rho}_{G(\mathfrak{p})} \otimes \mathfrak{h}^{\rho_2}_{G(\mathfrak{p})}$ is mapped onto the Hilbert space

$$(3.3.10)\quad A^{\rho\rho_2}(\mathfrak{h}^{\rho}_{G(\mathfrak{p})} \otimes \mathfrak{h}^{\rho_2}_{G(\mathfrak{p})}) = \oplus\int_{\hat{G}(\mathfrak{p})} \sqrt{d\tilde{\mu}_{\rho\rho_2}(\rho')}\,\mathfrak{h}^{\rho\rho_2}_{\rho'} \otimes \mathfrak{h}^{\rho'}_{G(\mathfrak{p})}$$

on which the product representation $U^{\rho}_{G(\mathfrak{p})} \otimes U^{\rho_2}_{G(\mathfrak{p})}$ decomposes into the direct integral

$$(3.3.11)\quad \oplus\int_{\hat{G}(\mathfrak{p})} d\tilde{\mu}_{\rho\rho_2}(\rho')\,(\mathbb{1}_{\mathfrak{h}^{\rho\rho_2}_{\rho'}} \otimes U^{\rho'}_{G(\mathfrak{p})})$$

of irreducible unitary representations $U^{\rho'}_{G(\mathfrak{p})}$ of $G(\mathfrak{p})$ each occurring with the multiplicity given by the dimension of $\mathfrak{h}^{\rho\rho_2}_{\rho'}$. The measure $\tilde{\mu}_{\rho\rho_2}$ on $\hat{G}(\mathfrak{p})$ again is uniquely defined up to equivalence by the decomposition. With the aid of the transformation $\oplus\int_{\hat{G}(\mathring{p})} d\tilde{\mu}_{\rho_1}(\rho)(\mathbb{1}_{\mathfrak{h}^{\rho_1}_{\rho}} \otimes \breve{A}^{\rho\rho_2})$ therefore the Hilbert space $\tilde{\mathfrak{h}}^{1,2}$ of (3.3.8) is mapped unitarily onto a Hilbert space which by a prescription similar to (3.3.7) may be identified with the Hilbert space

$$(3.3.12) \quad \hat{\mathfrak{h}}^{1,2} = \oplus\int_{\hat{G}(\mathring{p})} \sqrt{d\tilde{\mu}_{\rho_1}(\rho)} \oplus\int_{\hat{G}(\mathring{p})} d\tilde{\mu}_{\rho\rho_2}(\rho')(\mathfrak{h}^{\rho_1}_{\rho} \otimes \mathfrak{h}^{\rho\rho_2}_{\rho'} \otimes \mathfrak{h}^{\rho'}_{G(\mathring{p})}) .$$

If we carry over the representation $U^{1,2}$ from (3.3.9) to $\hat{\mathfrak{h}}^{1,2}$ we get

$$(\hat{U}^{1,2}(A,a)\hat{\psi})(p) := (U^{1,2}(A,a)\psi)^{\wedge}(p) =$$

$$= e^{ip\cdot a} \oplus\int_{\hat{G}(\mathring{p})} d\tilde{\mu}_{\rho_1}(\rho) \oplus\int_{\hat{G}(\mathring{p})} d\tilde{\mu}_{\rho\rho_2}(\rho')\Big[\mathbb{1}_{\mathfrak{h}^{\rho_1}_{\rho}\otimes\mathfrak{h}^{\rho\rho_2}_{\rho'}} \otimes \qquad (3.3.13)$$

$$\otimes U^{\rho'}_{G(\mathring{p})}(R(p;A))\Big]\hat{\psi}(\Lambda(A)^{-1}p) =$$

$$= \oplus\int_{\hat{G}(\mathring{p})} d\tilde{\mu}_{\rho_1}(\rho) \oplus\int_{\hat{G}(\mathring{p})} d\tilde{\mu}_{\rho\rho_2}(\rho')\Big[(\mathbb{1}_{\mathfrak{h}^{\rho_1}_{\rho}\otimes\mathfrak{h}^{\rho\rho_2}_{\rho'}} \otimes U^{\mathring{p},\rho'}(A,a))\hat{\psi}\Big](p),$$

i.e. on $\hat{\mathfrak{h}}^{1,2}$ the product representation $U^{1,2}$ of $\tilde{P}$ decomposes into a direct integral of multiples of irreducible unitary representations $U^{\mathring{p},\rho'}$ of $\tilde{P}$.

The two equivalence transformations needed above, $\breve{A}^{\rho_1}$ for the irreducible unitary representations of SL(2,ℂ) restricted to $G(\mathring{p})$ and $\breve{A}^{\rho\rho_2}$ for the product representations of $G(\mathring{p})$, partially can be found in the literature. For $\mathring{p} \in \{\pm m e_{(0)}: m > 0\}$, that is $G(\mathring{p})$ = SU(2), the reduction of $U^{\rho_1}_{SL(2,\mathbb{C})}|$ SU(2) is carried out in NEUMARKs book [18], while the reduction of the product representations of SU(2) leads to the well known Clebsch-Gordan series for SU(2) (cf. for instance BARGMANN [15]). For $\mathring{p} \in \{n e_{(3)}: n > 0\}$, i.e. $G(\mathring{p})$ = SU(1,1) the representations $U^{\rho_1}_{SL(2,\mathbb{C})}|$SU(1,1) were reduced in the papers of RÜHL [31], SCIARRINO and TOLLER [32] and MUKUNDA [33], while the Clebsch-Gordan coefficients for the product representations of SU(1,1) were calculated by HOLMAN III and BIEDENHARN Jr [34]. For $\mathring{p} \in \{\pm(e_{(0)}+e_{(3)})\}$, that is $G(\mathring{p})$ = E(2), this problem has been solved elsewhere (SCHAAF [37]).

For $\mathring{p}$ = 0 the product representation (3.3.1) has the form

$$(3.3.14) \qquad U^{1,2}(A,a)\psi = U^{\rho_1}_{SL(2,\mathbb{C})}(A) \otimes U^{\rho_2}_{SL(2,\mathbb{C})}(A)\psi \quad .$$

The reduction problem in this case is reduced to that for product representations of SL(2,ℂ). It was solved by NEUMARK [35] in a series of papers.

Appendix A: An Orthonormal Basis in $\mathcal{L}^2(\mathbb{R})$

We show that for fixed $\varkappa \in \{0,1\}$, $\ell \in \{0,1,2,\dots\}$ an orthonormal basis of $\mathcal{L}^2(\mathbb{R})$ is defined by

$$\text{(A.1)}\quad \begin{aligned} K_\mu^{\varkappa,\ell}(\lambda) :&= \frac{\Gamma(1+\ell+\varkappa/2-i\lambda)}{\sqrt{2\pi}}\, 2^{1+\ell+\varkappa/2} \left[\frac{(\mu+\ell+\varkappa)!}{(\mu-\ell-1)!}\right]^{1/2} \frac{1}{(2\ell+\varkappa+1)!} \times \\ &\times F(1+\ell-\mu, 1+\ell+\varkappa/2+i\lambda; 2\ell+\varkappa+2; 2), \quad \mu = \ell+1, \ell+2, \dots . \end{aligned}$$

Because of Gauß' relations between neighboured hypergeometric functions (ERDELYI [23], p. 103) for the $K_\mu^{\varkappa,\ell}$ holds the recursion relation

$$\text{(A.2)}\quad \begin{aligned} 2i\lambda K_\mu^{\varkappa,\ell}(\lambda) &= \sqrt{(\mu-\ell-1)(\mu+\ell+\varkappa)}\, K_{\mu-1}^{\varkappa,\ell}(\lambda) - \sqrt{(\mu-\ell)(\mu+\ell+\varkappa+1)}\, K_{\mu+1}^{\varkappa,\ell}(\lambda), \\ K_{\ell+1}^{\varkappa,\ell}(\lambda) &= \frac{\Gamma(1+\ell+\varkappa/2-i\lambda)}{\sqrt{2\pi(2\ell+\varkappa+1)!}}\, 2^{1+\ell+\varkappa/2}, \quad K_\ell^{\varkappa,\ell}(\lambda) := 0 . \end{aligned}$$

With the orthonormal basis

$$\text{(A.3)}\quad \left\{\psi'^{\varkappa,\ell,+}_\mu : \psi'^{\varkappa,\ell,+}_\mu(z) = \left[\frac{(\mu+\ell+\varkappa)!}{(\mu-\ell-1)!}\right]^{1/2} z^{\mu+\varkappa}, \quad \mu = \ell+1, \ell+2, \dots\right\}$$

of the Hilbert space $\mathfrak{H}'^{\varkappa,\ell,+}_{SU(1,1)}$ of functions holomorphic in the interior of the unit circle with a zero of order $\geq \ell+\varkappa+1$ at $z = 0$ with the scalar product

$$\text{(A.4)}\quad \langle\psi|\varphi\rangle' := \int_{|z|<1} \frac{dx\,dy}{\pi(2\ell+\varkappa)!}(1+|z|^2)^{2\ell+\varkappa}\, |z|^{-2\ell-2\varkappa-2}\, \psi(z)^*\, \varphi(z)$$

we form the generating function

$$\text{(A.5)}\quad w_{+,\lambda}^{\varkappa,\ell}(z) := \sum_{\mu=\ell+1}^{\infty} \psi'^{\varkappa,\ell,+}_\mu(z)\, K_\mu^{\varkappa,\ell}(\lambda)^* .$$

It obeys the differential equation

$$\text{(A.6)}\quad \begin{aligned} (1-z^2)\frac{d}{dz}\, w_{+,\lambda}^{\varkappa,\ell}(z) &= \left[(\ell+\varkappa+1)z^{-1} + (\ell+1)z + 2i\lambda\right] w_{+,\lambda}^{\varkappa,\ell}(z), \\ z^{-\ell-\varkappa-1}\, w_{+,\lambda}^{\varkappa,\ell}(z)\Big|_{z=0} &= \frac{\Gamma(1+\ell+\varkappa/2+i\lambda)}{\sqrt{2\pi}}\, 2^{1+\ell+\varkappa/2}, \end{aligned}$$

as is proved with the aid of the recursion relation (A.2). The solution is

$$\text{(A.7)}\quad w_{+,\lambda}^{\varkappa,\ell}(z) = \frac{\Gamma(1+\ell+\varkappa/2+i\lambda)}{\sqrt{2\pi}}\, z^{1+\ell+\varkappa} \left(\frac{1-z}{\sqrt{2}}\right)^{-1-\ell-\varkappa/2-i\lambda} \left(\frac{1+z}{\sqrt{2}}\right)^{-1-\ell-\varkappa/2+i\lambda} .$$

The integral $\int_{-\infty}^{+\infty} d\lambda\, w_{+,\lambda}^{\varkappa,\ell}(z')\, w_{+,\lambda}^{\varkappa,\ell}(z)^*$ essentially represents the Mellin transform of the absolute square of the Gamma function and yields the reproducing kernel (1.3.52) of $\mathfrak{H}'^{\varkappa,\ell,+}_{SU(1,1)}$:

$$\text{(A.8)}\quad \int_{-\infty}^{+\infty} d\lambda\, w_{+,\lambda}^{\varkappa,\ell}(z') w_{+,\lambda}^{\varkappa,\ell}(z)^* = K^{\varkappa,\ell,+}(z',z) = \sum_{\mu=\ell+1}^{\infty} \psi'^{\varkappa,\ell,+}_{\mu}(z')\psi'^{\varkappa,\ell,+}_{\mu}(z)^* .$$

By putting (A.5) into the left hand side and comparing the coefficients one obtains the orthonormality relations

$$\text{(A.9)}\qquad \int_{-\infty}^{+\infty} d\lambda\, K_{\mu'}^{\varkappa,\ell}(\lambda)^*\, K_{\mu}^{\varkappa,\ell}(\lambda) = \delta_{\mu'\mu} .$$

To proof the completeness of the system $\{K_{\mu}^{\varkappa,\ell} : \mu = \ell+1, \ell+2, \ldots\}$ in $\mathcal{L}^2(\mathbb{R})$ we show that a function which is orthogonal to all $K_{\mu}^{\varkappa,\ell}$ must vanish. The first parameter of the hypergeometric function in (A.1) takes negative integer values. Therefore the hypergeometric function is reduced to a polynomial of degree $\mu-\ell-1$ in λ that is connected with Pollaczeks polynomials (POLLACZEK [36]). Therefore it suffices to show that an element of $\mathcal{L}^2(\mathbb{R})$ that is orthogonal to all f_n: $f_n(\lambda) = \Gamma(1+\ell+\varkappa/2-i\lambda)\,\lambda^n$, $n = 0, 1, 2, \ldots$, must vanish. Let f be orthogonal to each f_n. Because of the asymptotic behaviour

$$\text{(A.10)}\qquad |\Gamma(1+\ell+\varkappa/2-i\lambda)|^2 \sim |\lambda|^{2\ell+\varkappa+1}\, e^{-\pi|\lambda|} , \quad |\lambda| \longrightarrow \infty ,$$

the function

$$\text{(A.11)}\qquad g_z : g_z(\lambda) = \Gamma(1+\ell+\varkappa/2-i\lambda)\, e^{iz\lambda}$$

for all z out of the strip $|\mathrm{Im}\, z| < \pi/2$ belongs to $\mathcal{L}^2(\mathbb{R})$. Then the scalar product

$$\text{(A.12)}\qquad F(z) := \int_{-\infty}^{+\infty} d\lambda\, f(\lambda)^*\, \Gamma(1+\ell+\varkappa/2-i\lambda)\, e^{iz\lambda}$$

exists for all z from this strip and represents a function which is holomorphic there. From our assumption about f follows

$$\text{(A.13)}\qquad \frac{d^n}{dz^n} F(0) = 0 , \quad n = 0, 1, 2, \ldots .$$

Then $F \equiv 0$ and therefore $f(\lambda) = 0$ for almost every λ.

Appendix B: The Sign Factor $\varepsilon(\omega,A)$

We discuss the phase factor

(B.1) $\varepsilon(\omega,A) := \omega^{-1/2}\zeta_A(\omega)(\omega\bar{A})^{1/2}$, $\zeta_A(\omega) := (\omega A_{12}+A_{22})/|\omega A_{12}+A_{22}|$,

from (2.5.53). Here $\omega\in\partial K$, $A\in SU(1,1)$. The square root is defined by

(B.2) $$\omega^{1/2} = e^{i\,\mathrm{arc}(\omega)/2}, \quad -\pi\leq \mathrm{arc}(\omega)<\pi .$$

Because of

(B.3) $$\varepsilon(\omega,A)^2 = \zeta_A(\omega)^2 \frac{\omega\bar{A}}{\omega} = 1 ,$$

$\varepsilon(\omega,A)$ only takes the values ± 1. With ω also $\omega\bar{A}$ runs through the boundary of the unit circle in the positive direction and takes the value -1 at

(B.4) $$\omega = \omega_- := -(A_{22}+A_{21})/(A_{11}+A_{12}), \quad \omega_- A = -1 .$$

According to (B.3) and (B.2) we have

(B.5) $$\mathrm{arc}(\omega\bar{A}) = \mathrm{arc}(\omega)-2\mathrm{arc}(\zeta_A(\omega))+\begin{cases} -2\pi \text{ for } \pi\leq \mathrm{arc}(\omega)-2\mathrm{arc}(\zeta_A(\omega)) , \\ 0 \text{ for } -\pi\leq \mathrm{arc}(\omega)-2\mathrm{arc}(\zeta_A(\omega))<\pi , \\ 2\pi \text{ for } \mathrm{arc}(\omega)-2\mathrm{arc}(\zeta_A(\omega)) < -\pi . \end{cases}$$

Since $\mathrm{arc}(\omega\bar{A})\geq \mathrm{arc}(\omega_-\bar{A}) = -\pi$ for $\mathrm{arc}(\omega)\geq \mathrm{arc}(\omega_-)$ from (B.5) follows

(B.6) $$\mathrm{arc}(\omega\bar{A}) = \mathrm{arc}(\omega)-2\mathrm{arc}(\zeta_A(\omega))+\begin{cases} -2\pi \text{ for } \pi\leq \mathrm{arc}(\omega)-2\mathrm{arc}(\zeta_A(\omega)) , \\ 0 \text{ for } -\pi\leq \mathrm{arc}(\omega)-2\mathrm{arc}(\zeta_A(\omega))<\pi . \end{cases}$$

If we let ω go to ω_- from the region $\pi > \mathrm{arc}(\omega)\geq \mathrm{arc}(\omega_-)$ both cases yield

(B.7) $$\mathrm{arc}(\omega_-) - 2\,\mathrm{arc}(\zeta_A(\omega_-)) = \begin{cases} +\pi , \\ -\pi . \end{cases}$$

Because of $-\pi\leq \mathrm{arc}(\omega_-)<\pi$ this corresponds to the alternative

$$(B.8)\qquad \begin{cases} -\pi \leq \operatorname{arc}(\varsigma_A(\omega_-)) < 0 \ , \\ 0 \leq \operatorname{arc}(\varsigma_A(\omega_-)) < \pi \ . \end{cases}$$

Therefore instead of (B.6) we can write

$$(B.9)\qquad \operatorname{arc}(\omega\bar{A}) = \operatorname{arc}(\omega) - 2\operatorname{arc}(\varsigma_A(\omega)) + \begin{cases} -2\pi & \text{for } -\pi \leq \operatorname{arc}(\varsigma_A(\omega_-)) < 0 \ , \\ 0 & \text{for } 0 \leq \operatorname{arc}(\varsigma_A(\omega_-)) < \pi \ , \end{cases}$$

if $\operatorname{arc}(\omega_-) \leq \operatorname{arc}(\omega) < \pi$. This completely fixes the change of sign of $\varepsilon(\omega,A)$, and we get according to (B.1)

$$(B.10)\qquad \varepsilon(\omega,A) = \begin{cases} +\operatorname{sign} \operatorname{Im}(A_{11}+A_{12}) & \text{for } \omega \in \overline{-1,\omega_-} \ , \\ -\operatorname{sign} \operatorname{Im}(A_{11}+A_{12}) & \text{for } \omega \in \overline{\omega_-,-1} \ . \end{cases}$$

Because of

$$(B.11)\qquad \begin{aligned} &\omega_\pm(\Gamma A) = -\omega_\pm(A), \ (\Gamma A)_{11} + (\Gamma A)_{12} = i(A_{11}+A_{12}) \ , \\ &\Gamma := i\sigma_3 = \begin{pmatrix} i & 0 \\ 0 & -i \end{pmatrix} \in SU(1,1) \ , \end{aligned}$$

from (B.10) follows

$$(B.12)\qquad \varepsilon(-\omega,\Gamma A) = \begin{cases} \operatorname{sign} \operatorname{Re}(A_{11}+A_{12}) & \text{for } \omega \in \overline{1,\omega_-} \ , \\ -\operatorname{sign} \operatorname{Re}(A_{11}+A_{12}) & \text{for } \omega \in \overline{\omega_-,1} \ . \end{cases}$$

With the upper of the relations

$$(B.13)\qquad 2\operatorname{Re}(A_{11}\pm A_{12})\operatorname{Im}(A_{11}\pm A_{12}) = \pm|A_{11}+A_{12}|^2 \operatorname{Im}\omega_\mp$$

we finally get

$$(B.14)\qquad \varepsilon(-\omega,\Gamma A)/\varepsilon(\omega,A) = \operatorname{sign} \operatorname{Im}\omega \ ,$$

which because of $\partial K_{\tau'\tau}(A) \subset \overline{\tau',-\tau'}$ can be written in the form

$$(B.15)\qquad \varepsilon(-\omega,\Gamma A)/\varepsilon(\omega,A) = \tau' \quad \text{for} \quad \omega \in \partial K_{\tau\tau}(A) \ .$$

Further because of

(B.16) $\omega_\pm(A\Gamma^{-1}) = \omega_\mp(A)$, $(A\Gamma^{-1})_{11}+(A\Gamma^{-1})_{12} = -i(A_{11}-A_{12})$

with (B.10) holds the relation

$$\varepsilon(\omega,A\Gamma^{-1}) = \begin{cases} -\text{sign Re}(A_{11}-A_{12}) & \text{for } \omega\in\overline{-1,\omega_+} \\ \text{sign Re}(A_{11}-A_{12}) & \text{for } \omega\in\overline{\omega_+,-1} \end{cases} \quad (B.17)$$

With the aid of the formulas

$$\text{Re}(A_{11}+A_{12})\,\text{Re}(A_{11}-A_{12}) - \text{Im}(A_{11}+A_{12})\,\text{Im}(A_{11}-A_{12}) =$$

$$= \tfrac{1}{2}\left|A_{11}{}^2-A_{12}{}^2\right|^2 \text{Re}(\omega_+-\omega_-) \quad , \quad (B.18)$$

$$\text{Re}(A_{11}+A_{12})\,\text{Re}(A_{11}-A_{12}) + \text{Im}(A_{11}+A_{12})\,\text{Im}(A_{11}-A_{12}) = 1 \quad ,$$

as well as with (B.13) one can show that the point -1 exactly lies in the arc $\overline{\omega_\tau,\omega_{-\tau}}$, if sign $\text{Re}(A_{11}-A_{12}) = -\tau$ sign $\text{Im}(A_{11}+A_{12})$. Since further the arc $\overline{\omega_\tau,\omega_{-\tau}}$ which contains the point -1 can be represented in the form $(\overline{-1,\omega_+}\cap\overline{-1,\omega_-})\cup(\overline{\omega_+,-1}\cap\overline{\omega_-,-1})$, with (B.10) and (B.17) follows

$$\varepsilon(\omega,A\Gamma^{-1})/\varepsilon(\omega,A) = \tau \quad \text{for } \omega\in\overline{\omega_\tau,\omega_{-\tau}} \quad , \quad (B.19)$$

or equivalently, since $\partial K_{\tau'\tau}(A)\subset\overline{\omega_\tau,\omega_{-\tau}}$,

$$\varepsilon(\omega,A\Gamma^{-1})/\varepsilon(\omega,A) = \tau \quad \text{for } \omega\in\partial K_{\tau'\tau}(A) \quad . \quad (B.20)$$

From (B.10) finally follows the symmetry relation

$$\varepsilon(1/\omega,A^*) = \varepsilon(\omega,A) \quad . \quad (B.21)$$

Literature

[1] H. JOOS, Fortschritte d. Physik 10, 65 (1962)

[2] R. MOUSSA, P. STORA in Brittin, Barut (Eds.), Lectures in Theoretical Physics VII A, Boulder: University of Colorado Press, 1965

[3] M. TOLLER, Nuovo Cim. 37, 631 (1965), 54A, 295 (1968)

[4] F.T. HADJEOANNOU, Nuovo Cim. 44, 185 (1966)

[5] H. JOOS in Brittin, Barut (Eds.), Lectures in Theoretical Physics VII A, Boulder: University of Colorado Press, 1965

[6] G. FEINBERG, Phys. Rev. 159, 1089 (1967)

[7] V. BARGMANN, Ann. of Math. 59, 1 (1954)

[8] E.P. WIGNER, Ann. of Math. 40, 149 (1939)

[9] G.W. MACKEY, Ann. of Math. 55, 101 (1952), 58, 193 (1953)

[10] G.W. MACKEY, Bull. Amer. Math. Soc. 69, 628 (1963)

[11] G. EMCH in Brittin, Barut (Eds.), Lectures in Theoretical Physics VII A, Boulder: University of Colorado Press, 1965

[12] J.C. GUILLOT, J.L. PETIT, Helv. Phys. Acta 39, 281 (1966)

[13] V. BARGMANN, Ann. of Math. 48, 568 (1947)

[14] I.M. GELFAND, M.I. GRAEV, N.Y. VILENKIN, Generalized Functions, Vol. 5, New York: Academic Press, 1966

[15] V. BARGMANN, Rev. Mod. Phys. 34, 829 (1962)

[16] R. TAKAHASHI, Jap. J. Math. 31, 55 (1961)

[17] I.M. GELFAND, M.A. NEUMARK, J. Phys. USSR 10, 93 (1946)

[18] M.A. NEUMARK, Lineare Darstellungen der Lorentzgruppe, Berlin: VEB Deutscher Verlag d. Wissenschaften, 1963

[19] N. MUKUNDA, J. Math. Phys. 8, 2210 (1967), 9, 417 (1968)

[20] N.Y. VILENKIN, Mat. Sb. 64, 497 (1964) (Russian), Am. Math. Soc. Transl. (2) 60, 159 (1967)

[21] HARISH-CHANDRA, Proc. Nat. Ac. Sc. 38, 337 (1952)

[22] I.M. GELFAND, R.A. MINLOS, Z.YA. SHAPIRO, Representations of the Rotation and Lorentz Group and their Applications, Oxford: Pergamon, 1963

[23] ERDELYI et al., Higher Transcendental Functions, Vol. 1, New York: McGraw-Hill, 1953

[24] I.M. GELFAND, G.E. SCHILOW, Verallgemeinerte Funktionen, Berlin: VEB Deutscher Verlag d. Wissenschaften, 1960

[25] S. HELGASON, Differential Geometry and Symmetric Spaces, New York: Academic Press, 1962

[26] ERDELYI et al., Higher Transcendental Functions, Vol. 2, New York: McGraw-Hill, 1953

[27] A. SOMMERFELD, Partielle Differentialpleichungen der Physik, 5. Aufl., Leipzig: Akadem. Verlagsges. Geest & Portig, 1962

[28] R.A. KUNZE, E.M. STEIN, Am. J. Math. 82, 1 (1960)

[29] E.C. TITCHMARSH, Introduction to the Theory of Fourier Integrals, 2nd Ed., Oxford: Clarendon Press, 1948

[30] J. DIXMIER, Les Algèbres d'Operateurs dans l'Espace Hilbertienne, Paris: Gauthiers-Villars, 1957

[31] W. RÜHL, Commun. Math. Phys. 6, 312 (1967)

[32] A. SCIARRINO, M. TOLLER, J. Math. Phys. 8, 1252 (1967)

[33] N. MUKUNDA, J. Math. Phys. 9, 50, 417 (1968)

[34] W.J. HOLMAN III, L.C. BIEDENHARN JR., Ann. of Phys. 47, 205 (1968)

[35] M.A. NEUMARK, Trudy Moskov Mat. Obšč. 8, 121 (1959), 9, 237 (1960), 10, 181 (1961) (Russian), Am. Math. Soc. Transl. (2) 36, 101 (1964)

[36] F. POLLACZEK, C. R. Acad. Sci. Paris 230, 1563 (1950)

[37] M. SCHAAF, Z. Physik 229, 336 (1969)

This series aims to report new developments in physical research and teaching – quickly, informally, and at a high level. The type of material considered for publication includes:

1. Preliminary drafts of original papers and monographs
2. Lectures on a new field, or presenting a new angle on a classical field
3. collections of seminar papers
4. Reports of meetings

Texts which are out of print but still in demand may also be considered if they fall within these categories.

The timeliness of a manuscript is more important than its form, which may be unfinished or tentative. Thus, in some instances, proofs may be merely outlined and results presented which have been or will later be published elsewhere.

Publication of *Lecture Notes* is intended as a service to the international physical community, in that a commercial publisher, Springer-Verlag, can offer a wider distribution to documents which would otherwise have a restricted readership. Once published and copyrighted, they can be documented in the scientific libraries.

Manuscripts

Manuscripts are reproduced by a photographic process; they must therefore be typed with extreme care. Symbols not on the typewriter should be inserted by hand in indelible black ink. Corrections to the typescript should be made by sticking the amended text over the old one, or by obliterating errors with white correcting fluid. The figures (in the original size) ready for reproduction should be inserted into the text. Should the text, or any part of it, have to be retyped, the author will be reimbursed upon publication of the volume. Authors receive 50 free copies.

The typescript is reduced slightly in size during reproduction, therefore a large size of type should be used; best results will not be obtained unless the text on any one page is kept within the overall limit of 18 x 26.5 cm (7 x 10½ inches). **The publishers will be pleased to supply on request special stationery with the typing area outlined.**

Manuscripts in English, German or French should be sent to Springer-Verlag, 6900 Heidelberg, Postfach 1780.

Die „*Lecture Notes*" sollen rasch und informell, aber auf hohem Niveau, über neue Entwicklungen in der Physik berichten. Zur Veröffentlichung kommen:

1. Vorläufige Fassungen von Originalarbeiten und Monographien.
2. Spezielle Vorlesungen über ein neues Gebiet oder ein klassisches Gebiet in neuer Betrachtungsweise.
3. Seminarausarbeitungen.
4. Vorträge von Tagungen.

Ferner kommen auch ältere vergriffene spezielle Vorlesungen, Seminare und Berichte in Frage, wenn nach ihnen eine anhaltende Nachfrage besteht.

Die Beiträge dürfen im Interesse einer größeren Aktualität durchaus den Charakter des Unfertigen und Vorläufigen haben. Sie brauchen Beweise unter Umständen nur zu skizzieren und dürfen auch Ergebnisse enthalten, die in ähnlicher Form schon erschienen sind oder später erscheinen sollen.

Die Herausgabe der „*Lecture Notes*" Serie durch den Springer-Verlag stellt eine Dienstleistung an die physikalischen Institute dar, indem der Springer-Verlag für ausreichende Lagerhaltung sorgt und einen großen internationalen Kreis von Interessenten erfassen kann. Durch Anzeigen in Fachzeitschriften, Aufnahme in Kataloge und durch Anmeldung zum Copyright sowie durch die Versendung von Besprechungsexemplaren wird eine lückenlose Dokumentation in den wissenschaftlichen Bibliotheken ermöglicht.